Advanced Materials
for Sports Equipment

ROBERT E. GREEN, JR.
MATERIALS SCIENCE & ENGINEERING
MARYLAND HALL
THE JOHNS HOPKINS UNIVERSITY
BALTIMORE, MARYLAND 21218

Advanced Materials for Sports Equipment

HOW ADVANCED MATERIALS HELP OPTIMIZE SPORTING PERFORMANCE AND MAKE SPORT SAFER

K. E. Easterling
School of Engineering
University of Exeter

CHAPMAN & HALL
London · Glasgow · New York · Tokyo · Melbourne · Madras

Published by Chapman & Hall, 2–6 Boundary Row, London SE1 8HN

Chapman & Hall, 2–6 Boundary Row, London SE1 8HN, UK

Blackie Academic & Professional, Wester Cleddens Road, Bishopbriggs, Glasgow G64 2NZ, UK

Chapman & Hall, 29 West 35th Street, New York NY10001, USA

Chapman & Hall Japan, Thomson Publishing Japan, Hirakawacho Nemoto Building, 6F, 1-7-11 Hirakawa-cho, Chiyoda-ku, Tokyo 102, Japan

Chapman & Hall Australia, Thomas Nelson Australia, 102 Dodds Street, South Melbourne, Victoria 3205, Australia

Chapman & Hall India, R. Seshadri, 32 Second Main Road, CIT East, Madras 600 035, India

First edition 1993

© 1993 K.E. Easterling

Typeset in 11/13pt Palatino by Acorn Bookwork, Salisbury
Printed in Great Britain by the University Press, Cambridge

ISBN 0 412 40120 7

A catalogue record for this book is available from the British Library

Library of Congress Cataloging-in-Publication data available

∞ Printed on permanent acid-free text paper, manufactured in accordance with the proposed ANSI/NISO Z 39.48-199X and ANSI Z 39.48-1984

This book is dedicated to my grandchildren:
Alexander, Simon and Christopher,
enthusiastic sportsmen all!

Contents

Preface

Sports and leisure activities have grown tremendously in recent years, to the extent that practically everyone enjoys one form of activity or another. Whether it be fishing, jogging, tennis, squash, skiing, cycling or boating, there is nowadays a very large selection of equipment available in the sports stores, with degrees of sophistication (and prices) to suit most people. Most equipment on sale today really is at the high-tech end of design and advanced materials construction, and, indeed, much of it derives directly from the aerospace industry itself! However, this does raise a problem for many of us. Since the cost per kilogram of most sporting goods is quite comparable with that of a jumbo jet (more in some cases!), how do we know what we are really purchasing? Also, is there really a need for this degree of sophistication since most of us are, after all, only 'weekend athletes'? This book attempts to elucidate the philosophy behind the design of sporting equipment and to explain how the use of these advanced space-age materials can both contribute to sporting performances and even help protect the participants from injury. In this respect, I deliberately keep to sporting equipment that is used exclusively under 'human power and endeavour' and omit the motor-power-assisted sports.

I have written this book for a very broad range of sports enthusiasts. It should be useful for the really dedicated, but equally I feel the weekend sportsmen and women will find much to enlighten them concerning equipment and its use and optimization. It should even interest the 'armchair-sportsman'; these people are sometimes ridiculed but amongst the elderly especially they are often most knowledgeable concerning aspects of sports that interest them. The book is written in a readable (not academic) form, but the facts and details are accurate and verified by expert referees. It should thus appeal as additional reading in courses in engineering design, materials science and engineering, bioengineering and sports medicine. In every case, I hope the reader finds the book enjoyable and useful.

Acknowledgements

In assembling the large amount of data and literature on the various sporting equipment discussed, I have had much help from my colleagues at Exeter and Lulea Universities. I would particularly like to single out Mr Patrick Kalaugher (Exeter) and Mr Lennart Wallstrom (Lulea). In addition, I have received useful comments and information from Dr Jack Harris (a leading UK consultant on materials) and Dr Hugh Casey (Los Alamos National Laboratory). I have also enjoyed enthusiastic support and assistance from Mr Marcus Kirsch (University of Hamburg) in the literature searches and initial manuscript layout, and excellent help from Solveig Nilsson (Lulea University) with some of the artwork.

Finally, I would like to thank my secretary, Mrs Jan Adkins (Exeter), for her patience and skill in typing and word processing when preparing the final manuscript.

Kenneth Easterling
Exeter, 1991

Sports equipment and philosophy of design

SYNOPSIS

The design and manufacture of sports equipment is based on several engineering and scientific disciplines. These include mechanical engineering design, biomechanics, fracture mechanics, materials science and many aspects of the anatomy and physiology of the participant. The design philosophy of sports equipment has nowadays two main objectives: to improve sporting performance; and to maintain freedom from injury. In this chapter we search for clues as to how these objectives may be achieved in practice.

INTRODUCTION

More than ever before, sport and leisure have become an integral part of our daily life. This is partly because most of us these days have more leisure time to spend as a result of shorter working weeks, a lower retirement age and a growing life expectancy; but it is also, however, a byproduct of changes in our lifestyle. Nowadays, regular physical activity is regarded as a means to keep fit and healthy. Physical exercise not only strengthens muscles, but also improves mobility and balance, increases stamina, keeps weight under control and (not to be forgotten) is often even fun! In other words, sport generally makes life healthier, brighter and more active. Indeed, in this day and age the popular image of the 'weekend athlete' as someone sitting in an armchair 'participating' in sports on TV (Figure 1.1) is no longer quite so true as it was perhaps.

In the United Kingdom alone, 21.5 million adults and 7 million children are estimated to participate regularly in some kind of sporting activity. Currently, people in the Western world alone spend billions of dollars per year on sporting goods. Indeed, present trends indicate that the percentage of the population participating in active sport or leisure is on the increase in all of the industrialized countries. This has obviously led to a boom in the sports and leisure industries since most people engaged in some kind of sporting activity appear keen to buy the latest

Figure 1.1 Active sport is gradually taking over from armchair sport . . . ?

equipment, whether it be simply a pair of reliable jogging shoes or the last word in carbon fibre composite fishing rods.

It is a curious fact that irrespective of whether it be a weekend jogger, or an Olympic champion, neither can resist having the very best equipment that money can buy. Highly specialized research laboratories run by international companies are only too well aware of this and are obviously willing to fulfil all expectations in this respect! The companies all invest a great deal of money in ensuring their products employ the latest techniques and materials; competition between the various concerns is fierce. Consequently, today's sports equipment really is the result of the latest in design concepts and advanced materials available, many of which are byproducts of the space age itself. Unfortunately this also has the effect of elevating the prices and while there is nothing wrong in sportsmen and women wanting to improve their performance – and be willing to pay for it – it is perhaps important for them to know what exactly they are getting for their money. This will be one of the aspects addressed in this book.

THE ROLE OF BIOENGINEERING IN SPORTS EQUIPMENT

Research in sporting goods is by no means restricted to its commercial aspects. The interface between medicine and engineering is nowadays quite apparent, and for very good reasons. The striking of a baseball, the storage of energy in a vaulting pole, the flight of a football thrown or kicked at high velocity (Figure 1.2) – all involved most of the basic medical and engineering disciplines, from the physiology of the participant to the sciences of applied mechanics, biomechanics, dynamics, kinematics, etc. In the production of sports equipment, a strictly interdisciplinary approach is thus necessary, not only to achieve maximum performance but, perhaps even more important when considering the greater part of the population, to achieve freedom from injury.

The growing number of people participating in leisure and sports activities has meant that research in the biomechanics of sports equipment has assumed considerable importance. In other words today's sports equipment is actually designed to **protect** the participant from injury and this will be a recurring theme in this book. Anyone involved with hard physical exercise is liable to endanger body or limb while training, and the risk of this is obviously greater for the weekend athlete. One of the key objectives when selecting materials for sports equipment is thus to minimize risk of injury. This may be achieved, for example, using shock-absorbing materials to lower possible adverse shock-loading effects on the bone structure or the joint ligaments concerned. Appropriate design and sound material selection can even help prevent injury when a slightly faulty technique is employed, a situation that is likely to occur only too often, as most part-time athletes will readily acknowledge! Typical complaints from incorrect sporting techniques include 'tennis elbow' and 'golfers' back', both of which can be alleviated by the right

Figure 1.2 Results of studies of the physiology of sporting activities are used to help design sports equipment and help avoid injury. From *Athletic Ability and the Anatomy of Motion* by R. Wirhead (Wolfe Medical Publications, 1989, p.125).

choice of material. We shall find that, practically without exception, the design of sports equipment of all types now takes into consideration this important aspect of protection of the user, and that in most cases this involves the subtle use of material composites in carefully chosen proportions.

ADVANCED MATERIALS IN THE DESIGN OF SPORTS EQUIPMENT

Optimization is really the key word in the production of today's sporting products. Materials are typically characterized by various mechanical property considerations such as strength, weight or mass, toughness, dampening behaviour, etc.; however, even aesthetic properties, and (of course) cost, are of importance. We shall be examining in detail these aspects of sports equipment in the next chapter.

In spite of the complexity of materials used in many applications, in terms of multi-component composition and resulting mixed properties, materials selected for a given application need to be amenable to forming and assembly in a production line. As an example of this, tennis balls are currently produced at the rate of 50 per minute; this has meant that the two hemispherical halves of the balls cannot be sewn as earlier, but that they need to be rapidly formed under pressure in a single machine operation. In other words sports equipment, even if it typically consists of a number of different materials and is likely to be of a fairly complex design, needs to be considered in the category of advanced, high-tech high-production components, just as in the production of advanced electronic equipment like TVs or telephones.

To meet the many demands of today's sports and leisure industry, we shall find that sporting goods often consist of a mixture of many different materials: metals, ceramics, polymers and (mainly) fibre-reinforced composites (Chapter 2). Indeed, the application of these advanced materials now directly affects practically every kind of sport: alpine and cross-country skis and sticks, tennis, squash and badminton rackets, running shoes, fishing rods, golf clubs, hockey sticks, boats, surfboards and

Figure 1.3 The huge increases in height achieved by leading pole vaulters depend to no small extent on the use of advanced materials. The latest poles employ complex composites of layered carbon fibre weaves (see Chapter 2).

masts, bicycle frames and wheels, canoes, sports shoes, etc. New materials are even being used in the construction of artificial playing surfaces and indoor and outdoor running lanes in which the use of cellular composites is helping to improve performance (by optimizing the springiness of the surface) and hence decrease wear or even injury to limbs. A couple of examples at this stage may help to elucidate some of the improvements in performance that have been obtained with the help of advanced materials.

An often-quoted example concerns the development of vaulting poles in recent years (Figure 1.3). In the 1960s the world record height for a pole vault was around 4.75 metres (about 15.5

feet). Today, it has exceeded 6 metres (about 20 feet). The difference obviously lies partly in the skill of the vaulter; however, it has to be admitted that the larger part of this improvement is actually due to advances in the types of materials used to make the pole. Thirty years ago, leading vaulters used the finest bamboo they could lay their hands on. In the 1960s the first light metal (aluminium) poles appeared and existing records immediately started to fall. The really giant leap (upward) for mankind, however, occurred when carbon-fibre-reinforced polymer (epoxy) poles appeared in the 1980s. These poles were not only lighter and stronger than earlier models but were custom built to possess just the right stiffness and springiness with respect to the weight of the champion vaulters to help them achieve ever giddier heights.

Another interesting application of advanced carbon-fibre-reinforced polymers in sports equipment concerns the design of fly rods, which, because of their lightness and strength, are able

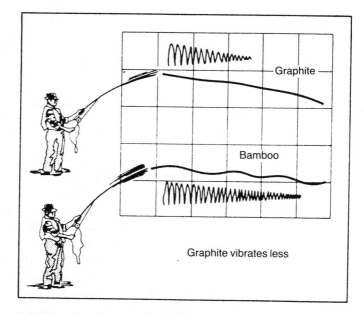

Figure 1.4 Vibrations in a graphite-reinforced epoxy fishing rod dampen more quickly than in a bamboo rod. The excessive vibration in bamboo wastes energy, whereas a graphite rod allows 20% more casting distance. Courtesy of Hercules.

to throw the flies considerably farther than the old bamboo rods. These advanced materials are also used in roach poles in which the old 9 metre (30 feet) long hollow bamboo poles have been replaced by two- or three-piece carbon-fibre-reinforced polymer rods (Figure 1.4), and at a fraction of their former weight. Not only are these new rods more comfortable to use, they are stronger and possess a better fatigue life than the old rods (see Box).

BOX

Advanced Composite Fishing Rods: how advanced materials improve a fisherman's skill

What decides how far a rod will cast, and (when baited) its sensitivity, is a combination of several different factors. High up in the requirements of the ideal fishing rod is **resilience**, a materials property that incorporates elasticity, toughness and strength. Another factor is **vibration**, which must be reduced to a minimum since any reverberation in the rod's shaft represents an energy loss which may steal many valuable metres from the length of casts. Casting distance is affected by other factors, too, of course. The number of line-runner rings and the material from which they are made are all important factors. In the latest technology these rings are coated on the inside with a hard ceramic such as aluminium oxide in order to minimize wear and friction to the running of the line. Yet other important factors to the property of a good rod are lightness and sensitivity or 'feel', properties that are highly dependent upon the materials used in the rod's 'backbone'. As shown in the illustration, modern rods are typically made up of a complex composite of fibre glass weaves, parallel carbon fibres and wound filaments of carbon fibre tape, each layer in the composite designed to account for, or optimize, a particular desired property of the rod. For example, the fibre glass/epoxy weave provides resilience and the parallel fibres oriented along the axis of the backbone provide great bending strength and elasticity to the rod. The outer carbon fibre weave reduces torsional strains and helps provide extra sensitivity to the rod. The whole, the advanced composite backbone of the rod, is thus a finely tuned light-weight instrument which (almost unfairly) presents the fisher-man with 'an edge' in his battle with the wiliest of fish.

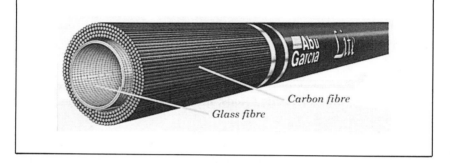

Carbon fibre

Glass fibre

THE SCIENCE OF BIOMECHANICS IN SPORTS

Biomechanics is a science that is concerned with the mechanical stresses and strains to which the human body may be exposed, whether during walking, running or participating in highly active sports like squash or football. In effect, it is an attempt to apply the basic laws of physics and mechanics to the joints and liga-ments and tissues of the body as they are subjected to both external loading and self-loading (Figure 1.5). Obviously, these laws of mechanics are not entirely accurate when applied to the human body because of the many individual variations in our physique, but useful guidelines can be obtained and these can be backed up by laboratory experiment and field trials. In any case, when considering the design philosophy of sports equipment it is useful to understand a little of the background of biomechanics applied to sports activities.

The main types of mechanical loading to which we can be subjected in sport are **tensile** (or pulling) loads, **compressive** loads, bending loads, **torsional** (or twisting) loads, and a some-what obscure but important form of load which can often lead to injury or fracture of limbs called a **shear** load (Figure 1.6). An example of shear loading is the cutting of paper by scissors. When designing and constructing sports equipment it is obviously necessary to do so in the context of the forces developed between our bodies and the particular type of mechanical loads of the

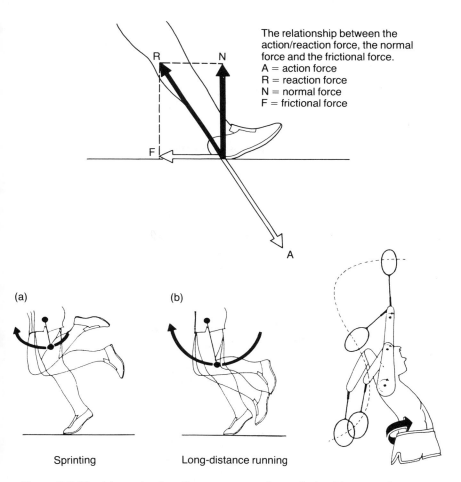

The relationship between the action/reaction force, the normal force and the frictional force.
A = action force
R = reaction force
N = normal force
F = frictional force

(a)

(b)

Sprinting   Long-distance running

Figure 1.5 The biomechanics of sport concern the analysis of forces and stresses which act on the body in the various sports. Studies of this sort contribute strongly in optimizing materials selection and equipment design. From *Athletic Ability and the Anatomy of Motion* by R. Wirhead (Wolfe Medical Publications, 1989, p.105).

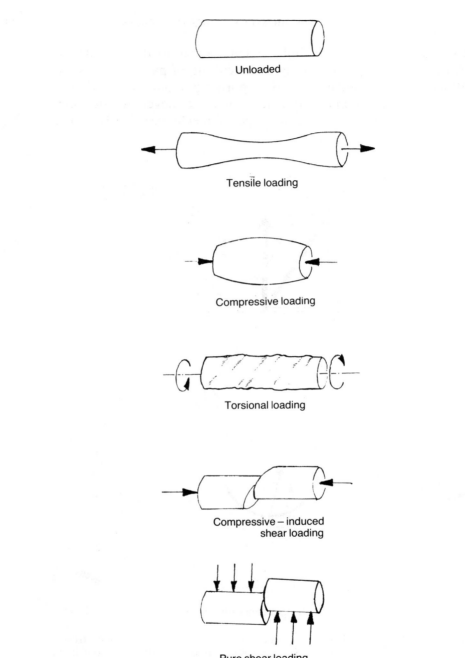

Figure 1.6 Schematic illustrations of some of the types of forces to which the body frame can be subjected.

sport concerned. Biomechanics is also a subject with a close relationship to anatomy, physiology and other medical practices such as physiotherapy and chiropractic manipulation. These studies, in cooperation with the producers of sports equipment, provide the basis of the growing appreciation and practice of safe (injury-free) sport.

CONCLUDING REMARKS

The philosophy behind the design and construction of sports equipment is concerned with many basic objectives and involves a number of scientific and medical disciplines. The new advanced materials used in sports equipment play a key role in several respects: they help improve performance; they must act as a buffer to the various types of loading, particularly shock loading, to which the body may be subjected; they must be aesthetically pleasing to the eye without loss or deterioration of their useful property; they may even provide extra 'feel' in sensitive sports such as fishing, golf, skiing, canoeing, etc. In this book we shall obviously emphasize the many various roles played by materials in sports equipment. However, we shall not forget that the design of sports equipment is essentially an interdisciplinary activity involving not only careful materials selection, but also many aspects of biomechanics, design, production engineering, fracture mechanics, materials science and, last but not least, the will and motivation of the participants concerned.

2 | Fundamentals of advanced materials

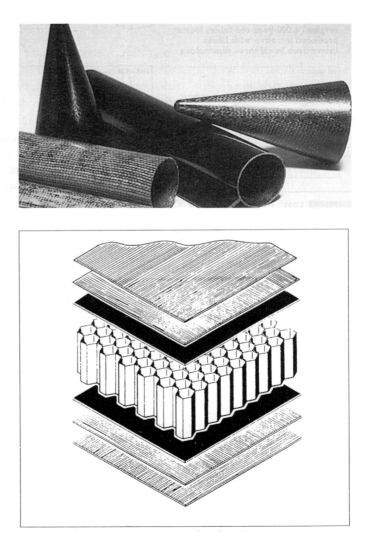

SYNOPSIS

The importance of advanced materials, such as fibre-reinforced composites, in providing the basis of strong, tough, lightweight as well as injury-proof sports equipment is now well established. In this chapter the special features of these materials, particularly composites, are examined in detail, including how they are made and why they possess the unique properties they do.

INTRODUCTION

Most applications of materials in everyday life are based on the use of relatively simple (single-component) solid materials, such as metals like steel or aluminium, or plastics. Ships, oil platforms, bridges and kitchen sinks, for example, are all made of different sorts of steels. Many building and household appliances, such as bowls, buckets, gutters, pipes, toys, and lavatory seats, are based on rather unsophisticated polymers (plastics). The houses themselves are mainly built from bricks and cement, while other household appliances such as cups, plates, lavatory bowls and bathtubs are made of rather brittle clay ceramics. By contrast, practically all sporting goods are nowadays based on complex and advanced forms of materials known as **fibre composites**. A composite is made up of two or more component solid materials (Figure 2.1). One component can be either in **particulate** or in **fibre form**; the other component is the binder of **matrix material**. As we shall now discuss, composites provide materials with a wider combination of properties than is available in single-component materials, and that this tends in particular to suit the rather stringent requirements of sports equipment.

COMPOSITE MATERIALS

The idea of composites is a simple one in that a material with one set of properties (e.g. a polymer) is mixed with another material of a different set of properties (e.g. glass or carbon fibres). By

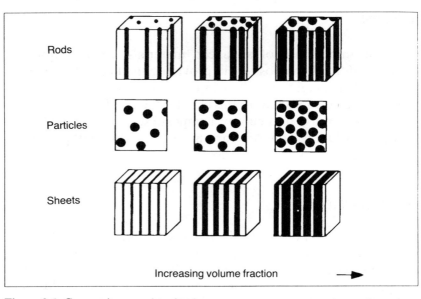

Rods

Particles

Sheets

Increasing volume fraction ➝

Figure 2.1 Composites consist of at least two components, or phases, in various proportions and shapes. From *Tomorrow's Materials* by K. E. Easterling (The Institute Materials, 1990).

mixing carefully, a composite of both materials can be produced with properties combining the best of both worlds. Examples of mixing two different materials together in various forms and proportions are shown in the illustration.

In terms of sports equipment, the properties of most importance may be listed as follows: lightness (low weight per unit volume), strength (the ability to withstand stresses and strains without breaking) (Figures 2.2 and 2.3), stiffness (the material's flexibility or resistance to bending or stretching) and toughness (the material's resistance to cracking even when exposed to sudden impacts). Other properties that can be important in sporting applications are dampening behaviour (this is in a way the opposite of stiffness and is the ability of a material to absorb rather than transfer impacting stresses to the human body), fatigue resistance (the material's resistance to cyclic stresses, e.g. constant impacts on a racket or club or on the ball itself), torsional resistance (the material's resistance to twisting, e.g. this has been a problem with golf clubs and vaulting poles). It will

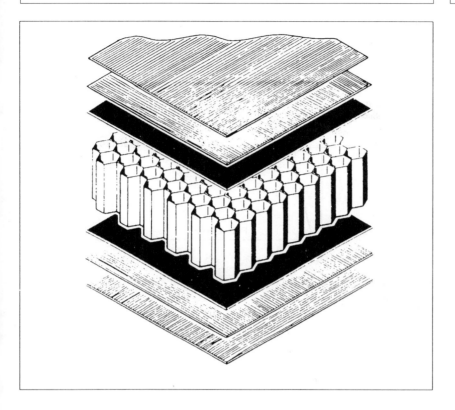

already be appreciated that no single material can be expected to possess all of these properties. Even the most advanced aluminium or titanium alloys can only cover certain of these specifications. A well-designed composite, however, may incorporate a whole set of desirable properties into the whole – the finished component – and is therefore much less restricted in applications than a single solid material.

RULE OF MIXTURES

In order to obtain a desired set of properties, materials scientists need to play a game of mix and match with two or more

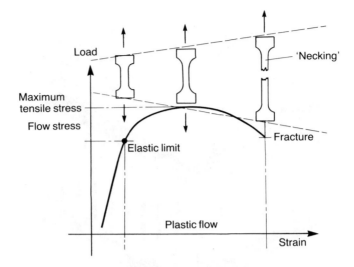

Figure 2.2 Definitions of strength, elastic limit and fracture. Stress is calculated from: load/sample cross-section.

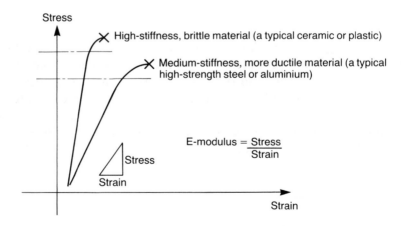

Figure 2.3 Illustrating that very stiff (high-modulus) materials are usually not very ductile (plastic).

component materials. Such mixing and matching yield a number of benefits. The simplest of these is known as the 'rule of mixtures'. This states, for example, that the strength of a composite is made up of the strength of its individual components in the proportions as they exist in the mixed material. Thus a composite that consists of 50% epoxy and 50% of chopped glass fibres is roughly 50% of the stiffness of epoxy and 50% the stiffness of glass. Since glass is stiffer and stronger than epoxy, whereas epoxy is more flexible than glass, the whole (the composite) is stiffer than epoxy and more flexible than glass.

By having one of the component materials in fibre form, many more advantages of a composite are made available. Indeed, clever design of fibre composites provides more than just a mixture of properties. For example, in a carbon-fibre-reinforced epoxy, the fibres can be laid in such a way that they are made to do most of the work in carrying the load along their length (Figure 2.4). Obviously, fibres are much stronger when loaded in tension along their length than when loaded from other directions. Thus the fibres can be oriented in such a way within the application of interest to provide strength or stiffness just where

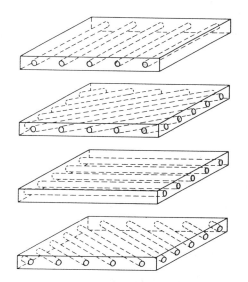

Figure 2.4 Illustrating that in composites the fibres can be oriented in a direction in which the highest strength is needed.

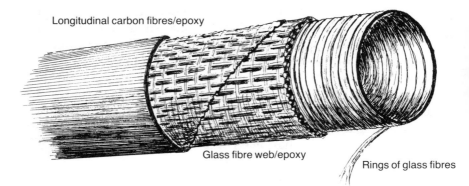

Longitudinal carbon fibres/epoxy

Glass fibre web/epoxy

Rings of glass fibres

Figure 2.5 A vaulting pole is made up of three layers of different fibres: an outer layer in which high-strength carbon fibres provide the pole with high stiffness, and an intermediate webbing of fibres together with an inner layer of wound glass fibres giving the pole resistance against twisting. Note that the glass fibre consists of 80% longitudinal and 20% radial fibres.

it is most needed. This can also be done so that the overall weight of the appliance is kept to a minimum. For example, in a vaulting pole, the fibres can be laid within the thin walls of the pole in such a way (Figure 2.5) as the minimize both excessive bending and twisting of the pole. This is achieved by arranging the fibres in double or treble spirals along the length of the hollow pole. In this case the epoxy resin binds the fibres together, but all of the applied work forces are taken by the fibres themselves. Obviously careful design, taking into account the external forces due to the vaulter, the density and orientation of the fibres, the thickness of the pole walls, etc., all help to optimize the final strength, flexibility and weight of the pole. In certain cases involving world-leading pole vaulters, companies are known to optimize their poles further by even taking into consideration the exact weight and height of the vaulter! However, this treatment is currently available for Olympic or world champions only.

CRACKING AND FRACTURE

When hard-working materials fail, they do so mainly by **cracking**. It takes a certain amount of energy to open up a crack in a material like a metal alloy or a ceramic, but once it gets going, or reaches some critical size, there is little to stop it growing. Unfortunately, the problem of cracking or fracture is not usually solved merely by using a stronger material, since it is nearly always found that the stronger a material, the lower its **critical crack size**. Thus advanced engineering ceramics like silicon nitride or alumina may need only a minute crack to form in them for the crack to expand through the material at breakneck speed. Indeed, a crack of only a fraction of a millimetre in length, when subject to some minimum (critical) stress, spreads in a brittle metal or ceramic with the speed of sound! Even cracks in hardened steel pipelines have been observed to grow tens of kilometres in a few seconds!

Composites are different in this respect, however. For example, in the Gulf War this was found to be an advantage in advanced military aircraft with fibre composite wings in that projectiles passed right through the composite without bringing about catastrophic spreading of fracture. All-metal wing structures may not have survived so well. The fibres absorb cracks (Figure 2.6) by allowing the crack to spread between the fibre and the epoxy glue. This crack-stopping effect is one of the great advantages of using composites in sports equipment and this

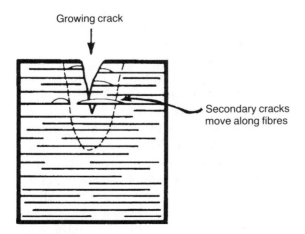

Figure 2.6 This illustration shows that a growing crack in a fibre composite is effectively stopped by the fibres. The secondary cracks along the fibres are relatively harmless.

feature brings out an important safety aspect too. For example, it is now standard competition practice to use carbon-fibre-reinforced epoxy squash rackets, instead of ones made from wood, since accidental smashes of the racket against the squash court wall can well cause injury from flying splinters. Composite rackets are built to absorb such impacts and thereby avoid splintering altogether.

THE RAW MATERIALS IN COMPOSITES: FIBRES AND BINDERS

The main types of fibres used in sports equipment are: carbon, glass, aramid or Kevlar, polyurethane (PU) and boron. Ceramic (silicon carbide or aluminium oxide) fibres are also now being used in certain applications. Of the matrix materials binding the fibres together, the most common are based on polymer resins such as epoxy or polyester. New tough binders are also being investigated and one of these is known as PEEK (polyetheretherketone).

The fibres

The most important fibre used in sports goods today is based on carbon. These fibres really consist of bundles of long chains of carbon atoms. They are made by burning an appropriate polymer fibre (polymers, or plastics, are all basically agglomerations of chemical chains); by stretching the fibre during the burning process a highly orientated – and hence strong – fibre of long chains of carbon atoms is produced. Under a powerful microscope, these chains are seen to be in the form of a ladder. The ladder-shaped chain, with its regular atomic cross-links, provides the carbon fibre (which is actually a thick bundle of intertwined 'ladder-chains') with its great strength and stability. Interestingly, one of the first uses of carbon fibres as a composite strengthener in sports equipment was in the shafts of golf clubs, golf being a sport in which enthusiasts are known to be willing to pay large premiums in exchange for a few extra metres down the fairway!

Fibre glass has been in use for many years, although even this

Table 2.1 Properties of various fibres used in composites

Property	Kevlar	E-glass	Carbon	SiC	Polyethylene
Mean fibre diameter, mm	2×10^{-4}	2×10^{-4}	10^{-4}	1.4×10^{-4}	5×10^{-6}
Density, Mg m^{-3}	1.44	2.54	1.8	3.0	0.97
Young's modulus, MN m^{-2}	1.24×10^5	7.2×10^4	2.2×10^5	4.0×10^4	1.2×10^5
Tensile strength, MN m^{-2}	2760	3450	2070	4830	2590
Elongation to facture, %	2.4	4.8	1.2		3.8

material is currently getting a new lease of life with the development of the high-grade, so called E-glass fibres (very high-stiffness fibres). Glass fibres are heavier than carbon fibres but they are still useful where cost, rather than weight, is the main consideration. An important area in sports equipment for glass-fibre-reinforced epoxy is in small boat and surfboard construction.

If toughness is a prime consideration, then aramid or Kevlar fibres are the best choice. These fibres are now found at the high-tech end of tennis racket production for example.

Ceramic fibres, such as silicon carbide and aluminium oxide, are also starting to be used in tennis rackets, where it is claimed they have better dampening (impact-absorbing) characteristics than the other fibres; they thus help to alleviate injuries such as tennis elbow. Some examples of fibres used in composites and their mechanical properties are shown in Table 2.1.

The binder

The matrix for accommodating and bonding the fibres together mainly consists of an organic polymer glue of epoxy. These polymer glues are viscous, rather sticky liquids which, with a suitable catalyst, are hardened to shape, usually aided by heat and pressure. The technique of production (Figure 2.7) is then to weave the fibres, or fabric of fibres, into the epoxy glue while it is still sticky, shape the whole into a preform, and then harden

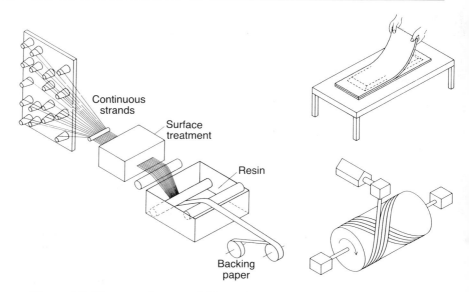

Continuous
strands

Surface
treatment

Resin

Backing
paper

Figure 2.7 The manufacture of fibre-reinforced polymer composites is illustrated in this figure, showing a continuous-strand unidirectional fibre composite and a spiral composite. The finished sheets of fibre composite are then laid out and pressure bonded. From *Scientific American* (October 1986, p.168).

it. Thermoset polymers such as epoxy and polyester are only pliable prior to being cured. When hardened, however, these epoxys are quite brittle. Bearing in mind that it is the fibres that do the work, not the matrix, brittleness does not really matter too much. Composites based on thermoset polymers are, with practice, relatively easy to make up yourself, and many enthusiasts construct their own boats or boards. Preforms are made by weaving, braiding, knitting or sewing fibres together (Figure 2.8) prior to shaping into a suitable moulded form. If highly directional properties are required, the preform can be made up of stacks of weaves of parallel fibres. The resemblance to the fibrous structure of wood cell walls is apparent at this stage of sophistication (Figure 2.9). Indeed the whole art of composites strives to emulate the ultimate perfection of growth and form that is nature. The epoxy (with catalyst) is finally impregnated into a preform, and the whole is moulded and baked in a large pressure cooker.

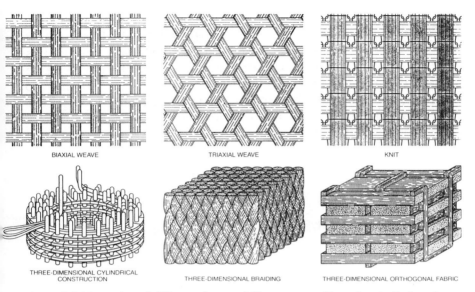

BIAXIAL WEAVE

TRIAXIAL WEAVE

KNIT

THREE-DIMENSIONAL CYLINDRICAL CONSTRUCTION

THREE-DIMENSIONAL BRAIDING

THREE-DIMENSIONAL ORTHOGONAL FABRIC

Figure 2.8 Examples of different forms of fibre weaves. Note the similarity of the three-dimensional cylindrical construction with the vaulting pole intermediate layer weave in Figure 2.5. From *Scientific American* (October 1986, p.168).

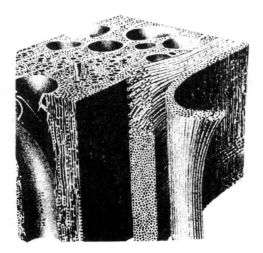

Figure 2.9 The layup of fibre cells in wood. Synthetic fibre composites strive to achieve the sophistication of nature in this respect. From *The International Book of Wood* (published by Mitchell Beazley, London, 1976, p.19).

A much tougher new binder currently being investigated in research laboratories is based on a polymer with the name of PEEK (polyether-etherketone). These tough polymer binders are much more expensive than ordinary epoxy, but are less prone to brittleness and cracking. Their current use is more confined to aerospace materials than sports equipment because of cost considerations. However, applications such as high-speed bobsleighs, or canoes required to negotiate rocky rapids, may well be able to utilize these superior tough polymers in the near future.

CELLULAR COMPOSITES

In many sports equipment applications, composites are in 'cellular' form. Again in striving to emulate nature, practically all natural forms are cellular. As already mentioned, wood is a typical example (Figure 2.10) of a natural cellular composite. If put under a microscope, wood is seen not to be 'solid' at all, but consists instead of stacks of open cells arranged together like bricks in a wall. Each cell is roughly of an elongated hollow hexagonal form, bound by walls of cellulose and stuck together by waxes and other glues. Within these finely symmetrically stacked cells are larger cells, and these are long tubes that carry moisture from the tree's roots to its outermost twigs. All trees have a similar microstructure; it is really only the cell wall thicknesses that distinguish one tree type from another. The cellulose in the wood cell walls consists of extremely fine fibrous sheets, bound together into a composite by suberine and other substances. Thus this composite cellular structure making up the tree is light in weight, has great stiffness (high elastic modulus) and is strong. This, of course, is the secret behind the impressive heights to which many trees grow and is why they can resist excessive bending even in windy weather. Indeed, when trees are felled by gale-force winds they usually fail by uprooting rather than by fracture of the trunk. Wood provides a model on which modern synthetic composite cellular structures are built (Figure 2.11). Modern skis and surfboards are both excellent examples

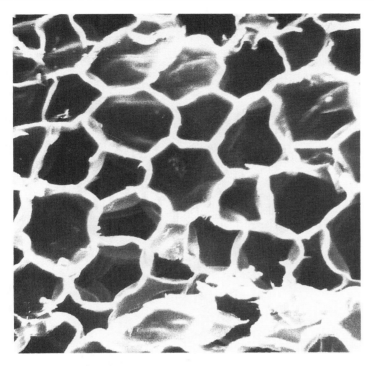

Figure 2.10 Illustrating the cellular structure of natural materials such as wood, cork and leaves. This photograph was taken in an electron microscope with a magnification of 10 000 diameters.

of such structures. In fact, these particular examples of sports equipment were, of course, originally made from wood.

COMPARING AND SELECTING MATERIALS

There are nowadays tens of thousands of different materials on the market. The question obviously arises as to how we can compare these many materials and then be able to select one of them because it has the best properties for the application we have in mind. This is no easy task since if you consult a catalogue listing different materials and their mechanical properties, it will typically only deal with one type of material. The main types of materials include:

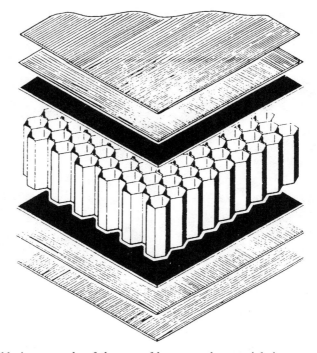

Figure 2.11 An example of the use of honeycomb materials in an engineering structure – part of an aircraft wing. These materials consist of a cellular structure of aluminium, polymer, or even paper, covered by one or more thin strong layers of a metal or fibre composite. The result is a lightweight material of very high stiffness, used in aircraft, skis and door panels. From *Tomorrow's Materials* by K. E. Easterling (The Institute of Materials, 1990, p.36).

- natural materials (e.g. rocks/minerals, wood, cork etc.)
- metals (including alloys)
- polymers (including elastics and foams)
- ceramics (both clay based and advanced engineering*)
- composites (various, including metal/ceramic, ceramic/ceramic, polymer/ceramic, polymer/metal, etc.)

A more detailed list of some of the forms of materials making up these various categories is given in Table 2.2.

*Advanced engineering ceramics (see Figure 2.12), sometimes referred to as fine ceramics, are made by sintering and densifying extremely fine powders from high-strength compounds such as silicon nitride, silicon carbide, aluminium oxide, boron nitride, etc. The fully dense material is very fine grained and very much purer and stronger than ordinary clay ceramics.

Table 2.2 Material types and some of their applications

Material type	Example*	Application*
Natural materials	Wood	Structural beams, houses
	Cork	Shoe soles, thermal and sound insulation
	Minerals and ores	Metal production, ceramics
	Sand (silica)	Glass, fine ceramics
	Rubber	High-grade rubber goods, tyres
Metals	Steel	Large structures (e.g. ships, bridges, oil platforms)
	Copper + alloys	Conductors, cables, bronze and brass items
	Aluminium + alloys	Conductors, structural profiles, aircraft skins
	Magnesium + alloys	Lightweight castings
	Silver	Photographic emulsion
	Gold	Jewellery
	Titanium + alloys	Aeroengine parts, chemical plant
Polymers	Polyethylene (PE)	Pails, pipes
	Polyvinylchloride (PVC)	Steering wheels
	Synthetic rubber	Tyres, conveyor belts, hoses
	Nylon	Clothing, racket strings
	Phenol formaldehyde	TV casings, telephones
	Terylene (Dacron)	Fibres and surface coatings
Ceramics	Diamond	Jewellery, indenters, electronics
	Alumina	Hard coatings, high-temperature components
	Silicon carbide	Hard fibres, wear-resistant coatings
	Aluminium nitride	Engine parts, coatings
	Boron nitride	Machine tools, coatings
	Zirconium oxide	Tools, engine components, coatings
Composites	Laminated wood	Beams
	Glass fibre/epoxy	Boats, skis
	Carbon fibre/epoxy	Skis, fishing rods, aircraft skins
	Kevlar fibre/epoxy	Surfboards, tennis rackets
	Silicon carbide fibre/ SIALON†	Refractory tiles

*This list is nowhere near exhaustive!
†A ceramic alloy of silica, alumina and silicon nitride.

Figure 2.12 The essential differences between the manufacture of conventional clay ceramics and advanced technical ceramics. Courtesy of *New Scientist.*

There are many different types of property we may be interested in. Some of these were introduced earlier in this chapter and included:

- strength (tensile, torsional, compressive, shear, etc.)
- lightness (weight per unit volume)
- stiffness (low deflection when loaded)
- fatigue resistance (materials can fail under repeated loading)
- dampening (particularly under impacting)
- fracture toughness (particularly against sudden failure)

As discussed earlier, in many sporting applications two or even more of these properties are invariably desired. In order then to achieve a satisfactory finished product composites need to be used in combination with clever design, taking into consideration biomechanical requirements to ensure freedom from injury.

Specific property

One way of approaching the problem of having to fulfil a number of different property requirements is to devise a method of comparing materials. This can be achieved in practice by comparing the **specific properties** of materials. For example, if we need to compare materials that have the requirement of both good strength and lightness then the specific property of interest is:

$$\frac{\text{strength } (\sigma_y)}{\text{weight per unit volume } (\varrho)}$$

i.e.

$$\text{specific strength} = \sigma_y/\varrho$$

In this formula, σ_y means the maximum strength that a material can reach before it permanently sets or deforms plastically; ϱ refers to the density of the material.

In a similar way, we could compare stiffness and lightness. This gives the equation

$$\text{specific modulus} = E/\varrho$$

where E is the modulus of elasticity; this modulus in effect describes the degree to which a material deforms **elastically** when loaded, and is thus related to the material's stiffness.

To illustrate this approach a number of different materials are compared in Table 2.3 in terms of their specific strength or their specific stiffness (E/ϱ).

Table 2.3 The role of design parameter on materials selection

Material	Young's modulus, E (MN m^{-2})	Density, ϱ (Mg m^{-3})	E/ϱ
Steel	210 000	7.8	26 923
Titanium	120 000	4.5	26 667
Aluminium	73 000	2.8	26 071
Magnesium	42 000	1.7	24 705
Glass	73 000	2.4	30 416
Brick	21 000	3.0	7 000
Concrete	15 000	2.5	6 000
Aluminium oxide	380 000	4.0	95 000
Silicon carbide	510 000	3.2	160 000
Silicon nitride	380 000	3.2	120 000
Boron nitride (cubic)	680 000	3.0	227 000
Boron fibres	400 000	2.5	160 000
Carbon fibres	410 000	2.2	186 000
Carbon fibre composite	200 000	2.0	100 000
Wood	14 000	0.5	28 000

Thus if a material is needed that gives the highest stiffness for the least possible weight, we select the material with the highest possible specific stiffness. As shown in Table 2.3 this material would be the advanced ceramic of boron nitride. If we decided that the material in question had to be metallic, then the best material would evidently be steel. This may sound a strange result when it is seen that the density of steel is nearly twice as high as that of titanium. However, this result tells us simply that if we needed a bar to have the highest stiffness for the minimum cross-sectional area, then steel is a better candidate than titanium, aluminium or magnesium. Note that if we use cellular materials instead of solid materials, the specific strength or stiffness in effect increases, since ϱ (cellular) is always less than σ (solid). This is in fact the basis of the design of much sports equipment such as skis, surfboards, ultralights, etc.

When designing complex sports equipment it is not usually as simple as described in our approach. We need instead to specify a more exact design criterion to bring about the optimum use of available materials.

In order to illustrate this point, it is useful to consider a couple of case studies. These case studies are obviously oversimplified, but they do give the general idea of the approach that can be used in practice. They may also assist the reader in appreciating why certain sports equipment is constructed in the way that it is.

CASE STUDIES IN MATERIALS SELECTION

Vaulting pole

We need to select the best material from which to construct a vaulting pole. Our specification could be as follows:

- We need the pole to be as light as possible.
- The pole must not buckle under the maximum load of the vaulter.
- The pole must not deform into any permanent set (no **plastic** strain is allowed).
- Loading is to be only compressive along the pole's axis.

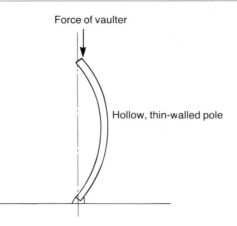

Force of vaulter

Hollow, thin-walled pole

Figure 2.13 The bending of a vaulting pole can be characterized in mechanical terms by a buckling of a long, slender column or tube. As such, the design criterion (see Figure 2.14) for maximum design stiffness is $\sqrt{E}/\varrho$.

- The pole needs to be stiff and only bend elastically (it must always return to its original shape).

These various constraints in our specification are illustrated in Figure 2.13. This type of problem is a common one in engineering design, and by consulting an appropriate reference book (see also Figure 2.14) we find that the specific design property is given by:

$$\text{specific property (max)} = \sqrt{E}/\varrho$$

where $\sqrt{E}$ refers to the square root of the elastic modulus.

Having obtained our specific design property, we must decide on which materials are of interest. We may decide to consider the following candidate materials:

- bamboo wood (this was used originally)
- aluminium (a light alloy)
- steel (strong and reliable)
- magnesium (a very light alloy)
- carbon fibre composite (very high strength and stiffness)

The appropriate values of $\sqrt{E}/\varrho$ for these materials are given in

Table 2.3. As before we need to select the material having the **maximum design stiffness**. We see that the best material is the carbon fibre composite, although (surprisingly) bamboo is not too far behind!

The next best after bamboo is the magnesium alloy, while the aluminium alloy and steel are apparently the least attractive

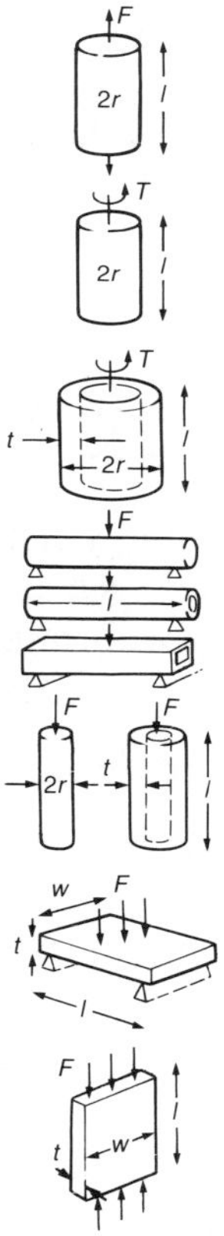

Mode of loading		Criterion to give minimum weight for given	
		Stiffness	Strength
Tie bar		$\dfrac{E}{\rho}$	$\dfrac{\sigma_y}{\rho}$
Torsion bar		$\dfrac{G^{1/2}}{\rho}$	$\dfrac{\sigma_y^{2/3}}{\rho}$
Torsion tube		$\dfrac{G^{1/2}}{\rho}$	$\dfrac{\sigma_y^{2/3}}{\rho}$
Bending of rods and tubes		$\dfrac{E^{1/2}}{\rho}$	$\dfrac{\sigma_y^{2/3}}{\rho}$
Buckling of slender column or tube		$\dfrac{E^{1/2}}{\rho}$	—
Bending of plate		$\dfrac{E^{1/3}}{\rho}$	$\dfrac{\sigma_y^{1/2}}{\rho}$
Buckling of plate		$\dfrac{E^{1/3}}{\rho}$	—

Figure 2.14 Examples of design criteria for various types of components and loading conditions. E refers to elastic modulus; ϱ refers to density of material; σ_y refers to yield strength; F refers to applied load; r is radius, t is thickness, l is length, W refers to width of plate and T is the torsional force.

materials for this application. World records are nowadays broken using a highly sophisticated form of carbon fibre composite hollow pole in which the design properties are optimized for specific stiffness, minimal torsion and ultralightness.

Bicycle construction

Design specification (see the figure in the appendix). The object of this case study is to select a suitable material for a bicycle taking into consideration the following criteria:

(a) the bicycle to be of the lowest possible weight;
(b) the bicycle to be built at the lowest possible cost using an acceptable (high-grade) material;
(c) possible materials: carbon fibre composites, steel, aluminium alloy, titanium alloy, magnesium alloy.

It is assumed that the frame components are of a tubular construction and may be simply represented as a cantilever beam as shown in Figure 2.15. It has been estimated* that the design criterion in this case is that the mass of the bicycle†, M, is given by:

$$M = \frac{2l^2}{3r^2} \frac{F}{\delta} \frac{\sigma}{E}$$

where these parameters are defined in Figure 2.15. The design criterion is thus given by:

$$M = \text{constant} \times (\varrho/E) = \text{a minimum}$$

the constant referring to that part of the equation which concerns fixed-value parameters.

The result is shown in Table 2.4 for the six materials listed. In this case, the best material is the one with the **highest specific design strength** (ϱ/E). It is seen that while the carbon fibre composite is the best material for top racing bicycles, the magnesium, aluminium and titanium alloys are also fairly satisfactory, while steel is almost as good as the light alloys. Of course, if

*See the appendix at the end of this chapter.
†Note that the **mass** of a body is not the same as its **weight**. If you stood on the Moon, your mass would be unchanged from its value on Earth; your weight, however, would be less.

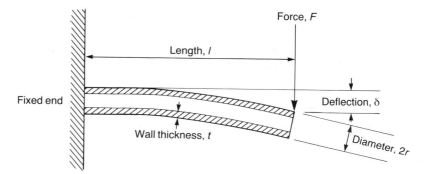

Figure 2.15 Simplified schematic of the elastic bending of a bicycle frame component such as a front fork.

Table 2.4 Materials for bicycles

Possible material	E^* (GN m^{-3})	ϱ^* (Mg m^{-3})	£† (10^2/ton)	$(\varrho/E \times 10^9$ (m^{-2} s^2)	£(ϱ/E) (£/GN m^{-1})
Steel	196	7.8	2.0	39.8	80
Wood	15	0.8	5.9	53.3	315
Aluminium alloy	69	2.9	10.0	39.1	391
Titanium alloy	120	4.5	50.0	37.5	1880
Magnesium alloy	42	1.7	30.0	40.5	1214
CFRP	270	1.8	900.0	5.6	5040

*Value based on actual bike materials.
†As of 1990.

overall cost (£ × ϱ/E) is taken into account (as it has to be in production models), steel is an obvious first choice since it possesses on the whole excellent properties at a minimum of cost. This is why most non-competition frames are still made from steel, of course.

APPENDIX

Case study of specific strength: bicycle construction

The object is to select a suitable material for a bicycle, taking into consideration (a) the lightest possible weight and (b) the lowest cost for an acceptable material.

Assume a tubular construction with a cantilever beam under bending as a critical member.

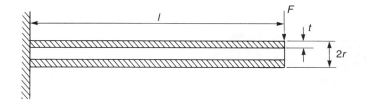

The deflection is

$$\delta = \frac{Fl^3}{3E\pi r^3 t}$$

i.e.

$$\frac{F}{\delta} = \frac{3\pi Er^3 t}{l^3}$$

The mass, M, is given by:

$$M = 2\pi rtl\rho$$

Substituting for t gives:

$$M = \frac{2l^4}{3r^2}\frac{F}{\delta}\frac{\rho}{E}$$

Hence

$$M = C\frac{\delta}{E} = \text{minimum}$$

3	# Sports shoes

SYNOPSIS

Human feet are extraordinarily complex biomechanical structures and as such highly prone to stress and injury. Jogging and running impose surprisingly large loads on feet, several times the weight of the body, while the transfer of load from one part of the foot to another is only accompanied via a quite complicated and seemingly unnatural route. All of this calls for sports shoe design and construction in which the use of advanced composite materials helps to combat the many forces of wear, tear, jar and sweat to which our feet and frames may be exposed.

INTRODUCTION

Human feet are complex biomechanical structures made up of 26 bones and about 30 joints, the whole being held together by a 'glue' of ligaments and muscles. To begin with it is worth considering the way this skeletal structure is loaded during running. This in turn provides clues as to how best to design sports shoes.

When running or jogging most people land on the heel, rotate to a midfoot position, and then take off via the ball of the foot. In terms of the foot's anatomy, two bones, the talus and the calcaneus (both part of the tarsus area) (Figure 3.1), are the parts of the foot first impacted. The load is then transmitted from the calcaneus to the upper parts of the lower extremities (the talus) as the running cycle is completed. the talus and calcaneus are themselves connected by a movable joint system called the sub-talar joint (Figure 3.2), and this construction allows for various movements to occur during walking or running about a relatively unstable axis. Movements of the foot about an unstable axis may sound strange, but most people exhibit this type of 'abnormality' in fact. The other tarsal bones, normally referred to as the midfoot, are by comparison very restricted in their movement; this is because there are a large number of ligaments that hold them tightly together. Normally, however, these last-mentioned bones do not undergo actual ground contact during running. Obviously, the foot's bones are also constructed to protect them-

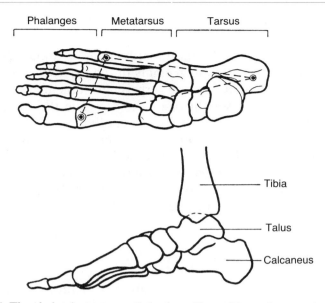

Figure 3.1 The skeletal structure of the foot. From *Biomechanics of Running Shoes* (ed. Benno N. Nigg, Human Kinetics Publishers, 1986, p.131).

selves and all of the major joints basically consist of two bones with a joint cavity or spacing between them. Each joint is embedded in a joint 'capsule' containing fluid in the cavity which acts as a shock absorber when impacted by the heel of the foot first hitting the ground.

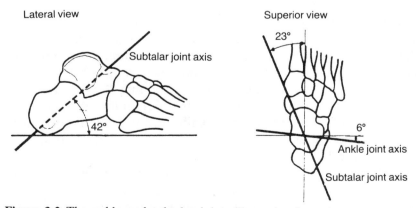

Figure 3.2 The ankle and subtalar joints illustrating how they are typically related by a fixed axis. From *Biomechanics of Running Shoes* (ed. Benno N. Nigg, Human Kinetics Publishers, 1986, p.133).

BASIS OF SPORTS SHOE DESIGN

In the construction of running shoes, the dynamics of the various forces to which the foot is subjected are obviously fairly complex. However, by using slow-motion photography to study athletes running on an indoor trainer a detailed picture of these forces can be built up. As mentioned above, it is found that 80% of all runners hit the ground on the centre heel (Figure 3.3), roll on to the midfoot and finally push off with the ball of the foot. Most of the remaining 20% actually land directly on the midfoot. Slow-motion photography further indicates that when the runner shifts the weight of the body from heel to midfoot, the foot rolls slightly inwards. This movement is called pronation.

During running the foot remains on the ground for about 0.2–0.3 seconds, dependent of course on the speed of the runner. It is this rather short contact time that sports shoe design is all about! In this fraction of a second, the foot needs to be cushioned from damaging impact forces, guided and stabilized in the complex movement as the body's load is transferred from the heel to the ball of the foot, while (ideally) still returning most of the runner's input energy in the form of a rebounding energy when the sole is compressed (this is called 'energy return'). Several parameters thus affect the heel-to-toe movement in running; these include running speed, road or track surface and (last but not least) design and construction of the shoes themselves.

In a normal run, an athlete takes about 300–400 steps per kilometre. At each step, the impact force on his or her heel is about two to three times the body weight. Thus when selecting shoe materials, they will need to possess good resilience against these repeated loads, in addition to the other basic requirements mentioned above. Yet another factor to be considered is the fact that no two people have exactly the same-shaped feet. In many cases these differences can be substantial; indeed, most people should ideally wear sports shoes that are tailor-made just for them.

ATHLETES' FEET

Every foot is unique, as is the person to whom it belongs. Indeed, an average foot is actually more or less 'deformed', a fact that

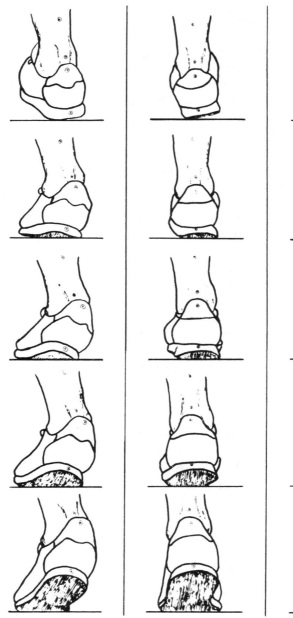

makes the construction of running shoes only more difficult (Figure 3.4). A comparison of footprints (podograms) well illustrates this.

The three extremes of shape are: the normal or unobtrusive foot; the flatfoot (pes planus); and the high arch foot (pes canus). The two latter types reveal a flattening or a raising of 'two-arch' construction inside the foot. High arch feet are sometimes linked with valgus or varus positions, as a result of rearfoot instability (Figure 3.5). In each case, running shoe design and the use of suitable materials need to negate these anomalies if injury is to be avoided.

Two main reactions are to be observed: overpronation and oversupination (Figure 3.6). While most heel strikes pronate to some extent, people with low arches tend to misalign feet by over 10°, from the normal to the axis of the shin (lower leg). This excessive pronation (or overpronation) increases the load on the foot because it results in an increased rotation of the foot, causing the knee and the hip joints to be subjected to an increased load. This may bring about problems known in the orthopaedic profession as injuries to the patellae, the tibia and the bursa. These problems may be avoided in sports shoe construction by splitting the midsole into softer and harder zones (Figure 3.7) producing double density soles.

A minority of runners perform a more outward-turned rolling movement from heel to toe. This is referred to as oversupination (also called off-supination) (Figure 3.8). In this case, the largest part of the impact force on the foot impacts the forefoot and the outer border of the foot. Oversupination is actually regarded as the cause of Achilles' tendon pain. It may also influence the whole running performance negatively because it has the effect of hindering natural forward movement of the body.

Figure 3.3 The rotary position and movement in the ankle joint for three runners representing an overpronation (left), very little pronation (centre), and also very little pronation but always remaining on the outside of the foot (right). All examples are of the left foot for heel-strike on the top and for intervals of about 50 m s^{-1} from top to bottom. From *Biomechanics of Running Shoes* (ed. Benno N. Nigg, Human Kinetics Publishers, 1986, p.7).

FOOT TYPE	HEIGHT OF LONGITUDINAL ARCH	REARFOOT POSITION	PODOGRAM
Normal	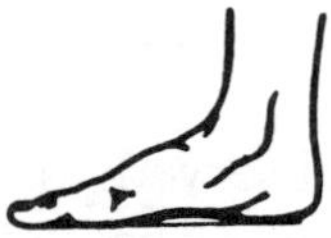	 0–6°	
Pes Planus			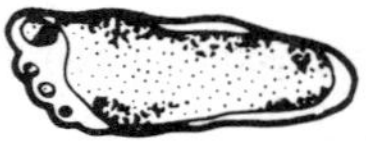
Pes Cavus			
Pes Valgus		 6° and more	
Pes Varus		 less than 0°	

Figure 3.4 Podograms of the shapes and distortions of feet. From *Biomechanics of Running Shoes* (ed. Benno N. Nigg, Human Kinetics Publishers, 1986, p.130).

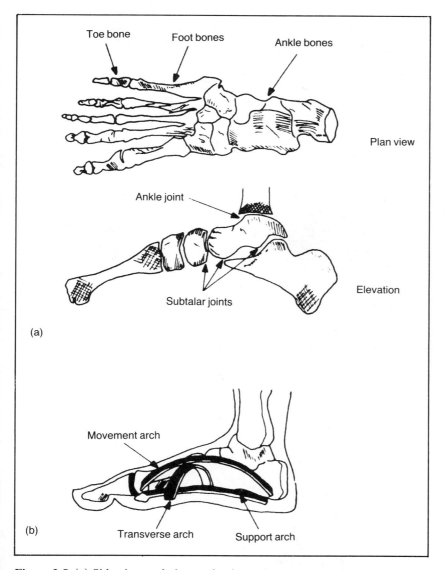

Figure 3.5 (a) Side view and plan projections of the bones of a human foot. (b) The foot represented in terms of its three support arches. Modern sports shoe design tries to compensate against overstressing these arches. Adapted from *Athletic Ability and the Anatomy of Motion* by R. Wirhead (Wolfe Medical Publications, 1989, p.59).

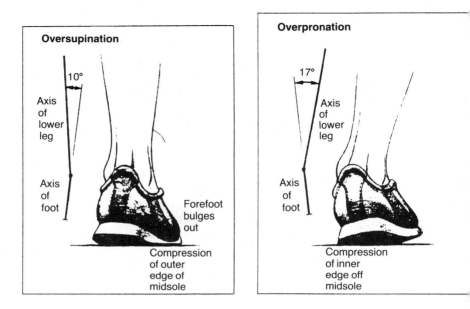

Figure 3.6 Illustrating how, as the runner's weight shifts from the heel to the midfoot area, the foot tends to roll inward, or pronate. In some cases (the minority of runners) the foot rolls outward, or supinate. Both cases need shoes that compensate for their particular style.

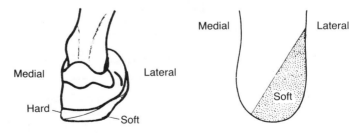

Figure 3.7 Possible solution of shoe which addresses the requirements of reduction of pronation using a double-density sole in the rear part as well as a slightly rounded lateral heel. From *Biomechanics of Running Shoes* (ed. Benno N. Nigg, Human Kinetics Publishers, 1987, p.155).

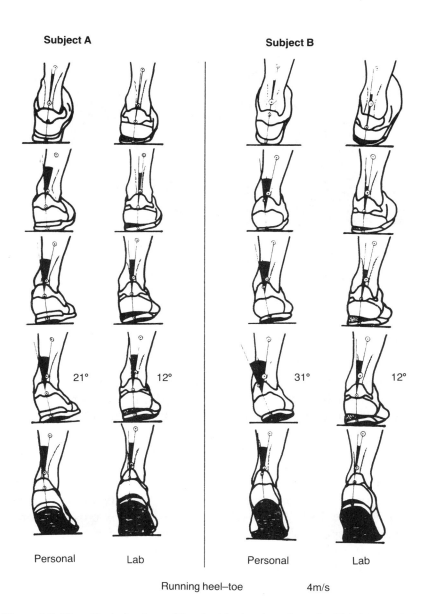

Figure 3.8 The effect of a shoe with a double-density sole and a slightly rounded heel for two runners which had excessive pronation in their commonly used shoes. From *Biomechanics of Running Shoes* (ed. Benno N. Nigg, Human Kinetics Publishers, 1987, p.157).

Figure 3.9 Various parts of a running shoe. From *Biomechanics of Running Shoes* (ed. Benno N. Nigg, Human Kinetics Publishers, 1986, p.119).

SPORTS SHOE ANATOMY

By now it should be realized that the number of parameters involved in running shoe design and construction is formidable. Apart from the runner's weight, sex, body structure, etc., we need also to consider individual foot type, typical mileages run and running style. When all of these parametres are taken into account running shoes need to consist of 15–20 individual parts and incorporate a wide array of different materials. The materials actually used are selected according to their resilience, strength, elasticity (stretchability), compression, durability and wear resistance (Figure 3.9). The resulting rather complex cross-section of a modern sports shoe is a model of advanced high-technology design. It is worth considering some of the individual components making up a sports shoe in more detail.

THE MIDSOLE

The key component of a sports shoe is its midsole, which is located between the upper and the outsole. The material of this

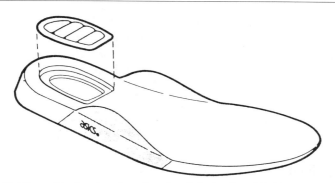

Figure 3.10 The midsole, showing also the Asics' gel pad insert.

component is compressed and tensioned many times during running and jogging. Basically it should provide cushioning and stability, in a way that does not diminish the shoe's energy return (Figure 3.10). To some extent this is a contradiction in itself, and it is a problem that can only be solved by clever design and materials selection.

Most sports shoe midsoles are made of EVA, a polymer or plastic foam consisting of ethylene, vinyl and acetate. Each of these ingredients has a different function: ethylene provides mouldability, vinyl gives resilience and acetate represents the part with strength and stiffness. These ingredients are mixed with a catalyst and poured into a mould. During the polymerization reaction that follows millions of tiny gas bubbles are generated which expand and fill the mould. The chemical composition, the molecular structure and the number of bubbles can, in effect, be varied to modify the material's hardness and strength as well as its weight and resilience.

Unfortunately, the walls of the air bubbles may break down after a hundred kilometres or so of service, and this may bring about a substantial reduction in cushioning effect. Some sports shoe manufacturers are trying to solve this problem by packing the bubbles closer to the surface using a special compression moulding technique (e.g. Etonic's Europa). Other companies try to bring about changes in density of the EVA by forming it as a tough 'integral' skin with a softer (less dense) foam at the outside (e.g. Puma's TX3).

Several running shoe companies have come up with an innovative solution to this problem of cushioning. In keeping with the theme that different parts of the foot ideally require different treatment, the area under the heel and ball is supported by soft EVA cushioning, while the rearfoot area of the sole consists of a firmer (higher-density) EVA for pronation control. Finally, the heel wedge consists of hard (fully dense) EVA (e.g. Reebok's Phase I Trainer). Other companies have tried using alternative pillars of high- and low-density foam (e.g. Saucony's The Flite). Asics Alliance on the other hand use a dense layer of EVA glued to a layer of porous EVA with a 'ripple' interface.

In addition to these variations, a crescent-shaped fibre glass plate can be sandwiched between the midsole and the heel wedge in order to bring about the desired properties (e.g. Etonic's Quasar). Basically all of these solutions attempt to reach a compromise between the several competing requirements of adequate cushioning, flexibility, stiffness and good rebound properties.

Encapsulated midsoles

This type of midsole construction is generally regarded as the most sophisticated attempt yet to combine adequate cushioning with good energy of rebound. Here the midsole material consists of polyurethane (PU) foam containing encapsulated layers of various other materials. The manufacturing process to achieve this is similar to that of EVA. Two liquid chemicals, polyol and isoxyanate, acting as resin and catalyst, are first mixed together, and during the resulting polymerization reaction gas bubbles are generated within the material. It should, however, be pointed out that while PU foam is more resilient, it is also more expensive. Perhaps the advantage of this material lies in its ability to encapsulate other materials.

A number of companies have now introduced midsoles in which inlets of EVA or polyethylene (PE) foam are embedded in PU (e.g. New Balance's 1300). Indeed, this clever innovation has resulted in the best of both worlds based on the combination of EVA's good cushioning properties with PU's excellent durability. The Puma R-System goes even further in providing three different layers of PE foam, all encapsulated in PU: the soft foam in the heel is for cushioning, the medium foam is located at the instep, and the hard foam is located along the inside as a pronation control component. Reebok's Energy Return system, on the other hand, encapsulates hollow tubes of Du Pont's Hytrel within a PU midsole. Hytrel is a very flexible thermoplastic which quickly springs back to its original shape when compressed and released.

With the invention of the Air Max in 1984, Nike was the first to use 'air pockets' to provide protection from impact force during running (Figure 3.11). These inlets are in fact not filled with air but with a pressurized gas giving lightness and cushioning, the whole being encapsulated in PU. Additionally, the midsole may be split into two parts: a high-density PU around the heel, with a lower density in the rest of the sole. Hi-Tec's ABC (Air Ball Concept) uses a similar technique.

It should be pointed out that critics of these various advanced

Figure 3.11 Cross-section from a 'Nike Air' shoe.

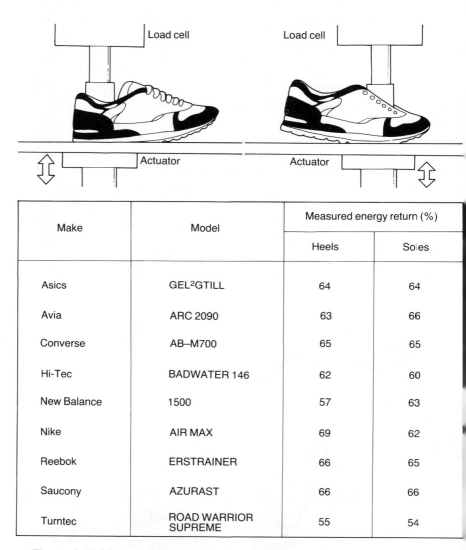

Make	Model	Measured energy return (%)	
		Heels	Soles
Asics	GEL²GTILL	64	64
Avia	ARC 2090	63	66
Converse	AB–M700	65	65
Hi-Tec	BADWATER 146	62	60
New Balance	1500	57	63
Nike	AIR MAX	69	62
Reebok	ERSTRAINER	66	65
Saucony	AZURAST	66	66
Turntec	ROAD WARRIOR SUPREME	55	54

Figure 3.12 Measurements of 'energy return' of various sports shoes (i.e. percentage of work of deformation got out relative to that put in) using the method shown. The general conclusion is that there is relatively little difference between shoes. After R. M. Alexander and M. Bennett (1969) How elastic is a running shoe?, *New Scientist*, 15 July.

designs claim that energy-return systems are being developed at the expense of safety and stability, and with scant thought of cost to the consumer. Indeed, this scepticism has been substantiated to some extent by some control tests (Figure 3.12) on a number of different makes, which revealed very little difference in the amount of energy return irrespective of make.

Obviously other approaches can be made in shoe construction which also provide support to the foot during running. Reebok's Pump Technology allows the wearer to inflate the lining by a device located in the tongue of the shoe. Brook's Hydro Flow System contains a two-chambered bag of silicon fluid in the midsole; during running this fluid is forced from one chamber to the other.

Some new research currently being carried out goes in another direction. The Adidas Torsion System, for example, attempts to compensate for rotation of the foot along the foot's longitudinal axis, with the goal of achieving an independent forefoot and heel movement. Two elements are necessary to bring this about: a groove in the sole, and a 'torsion' bar (reinforced with Kevlar) rotating along the axis, thus restricting lateral movement. Adidas has also developed the Delinger Web, a nylon netting wrapped around the top and the sides of the midsole to prevent it from deflecting and bulging out.

THE OUTSOLE

The outsole (Figure 3.13) is the part of the sports shoe that is in contact with the running surface; the two required properties are

Figure 3.13 The outsole and treads.

thus wear and frictional resistance. The most widely used mate-rial for this component is a styrene butadene rubber. A yet more abrasion-resistant rubber has the name of Vibram's Infinity, a high-volume fraction carbon fibre compound used in racing-car tyres!

Another approach for improving abrasion resistance is based on the so-called 'blown' technique. As with methods developed for generating PU bubbles, the material in this technique is expanded in a mould and then die cast. The materials used for this are based on styrene butadene or carboxylated nitryl, both of which have properties of lightness, resilience and good cushioning. A problem with these materials, however is their durability which is rather poor. On this basis, a compromise solution has been tried consisting of solid rubber pads cemented or moulded to a blown rubber outsole.

Treads

Tread patterns available on the market are as numerous as the number of different types of shoes themselves. Basically, these treads, or studs, need to provide traction and shock absorption, but they must also provide a 'good grip' to the shoe. Nike, for example, has developed a waffle pattern in their Odyssey Shoe that places waffle studs around the heel. During heel strike they behave like support columns which help to distribute the impact forces as well as dissipate the shock itself. Another functional tread pattern consists of 'sloped' triangular studs, which are compressed and pushed into the EVA midsole for additional cushioning (e.g. Saucony's Jazz). New Balance has come up with a design called the Hound's Tooth – a series of studs linked together and claimed to provide really good traction at the surface.

Adidas has equipped their outsole with small protusions which cover the sole and these are designed to behave as tiny shock absorbers. It is claimed that each of these may be compressed independently, according to the distribution of forces acting, thus providing an optimum and comfortable pressure on the foot.

THE INSOLE

Most insoles in today's running shoes consist of moulded PE foam
with a laminated fabric cover (Figure 3.14). Its function is to
achieve good fit and comfort, to prevent blisters and to absorb
impact shocks during the heel–toe movement. Since this is the
part of the shoe that is directly in contact with the foot itself,
those with special foot problems may also need orthopodic pads
to achieve optimal fit. Depending on their function some special
insole constructions employ quite different materials solutions.
For example, a layer of soft PU provided for cushioning with a
denser PU foam placed under the heel for additional support, is
a solution used by New Balance's 1300. Other solutions resort to
built-in supports in the form of an arch support under the insole
which may if necessary be exchanged for other orthopodic
devices.

Figure 3.14 The insole.

THE HEEL COUNTER

The heel counter supports the heel of the foot during the first part
of the heel–toe movement. In most cases the heel counter (Figure
3.15) is made of a light compressible material in order to achieve
the desired cushioning effect, although it is important that the
heel is also guided and supported. For this application, therefore,
rigid, durable materials need to be used. This problem is normally solved by incorporating a sheet of thermoplastic in the
counter which may also be reinforced by a high-strength
injection-moulded plastic in order to obtain better stiffness.

Figure 3.15 The heel counter.

External counter stabilizers are also used to minimize rearfoot instability. The external stabilizer may even be extended along the inner side of the shoe for overpronation control (e.g. the Adidas ZX 500). Another approach used is to reduce the problem of rearfoot instability by employing firmer midsole foams, or by integrating the heel counter and the midsole into a single unit. Converse and Asics have developed analogous systems in order to integrate the heel counter more closely with the midsole. Converse fits the heel counter with two moulded rubber flanges (stability bars), while Asics extends a layer of midsole foam to the heel counter.

THE UPPER

Upper materials are designed to achieve strength with minimal weight. Most constructions consist therefore of three layers (Figure 3.16): an outer PU foam, and a sheet of nylon tricot or

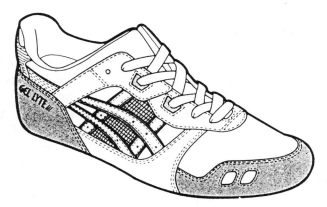

Figure 3.16 The upper including lacing system.

non-woven Canbrelle. Because of these materials' lack of poros-
ity, however, adequate ventilation of the foot may be somewhat
restricted and this leads to increased perspiration. Many companies thus prefer weblike layers of adhesive to bond the trilam-
inate together. Obviously, all manufacturers would prefer to use
quite new materials that not only are strong but also cope with
perspiration; the possibilities tried include synthetic leathers,
special blends of polyurethane and polyester, polypropylene
fabrics or Gore-tex, etc. Indeed, in very high-tech designs, the
upper alone may consist of as many as 20 separate parts, all of
which have to be sewn together and attached to the sole construction. Two methods currently used for connecting the upper to the
sole are known as 'board lasting' and 'slip lasting'.

When board lasting, a piece of fibreboard is first stapled to the
last. The upper is then pulled over it and cemented to the board
which provides a base for the midsole and the outsole. These
shoes have excellent stability but rather limited flexibility. The
slip lasting method, on the other hand, requires no board, and in
this case the upper is stitched together and fitted on to the last.
While these shoes are rather flexible, they lack somewhat in
overall stability. The best solution should obviously lie in a
combination of both methods. This is now available and is known
as 'combination lasting': in this approach the rear part is board
lasted, while the forefoot area is slip lasted. This thus provides
both rigidity in the area around the ball and flexibility in the heel
area.

FURTHER DOWN THE ROAD

The various examples presented above show how the manufacture of running shoes has become a sophisticated technology
requiring in turn a highly mechanized and computerized production process (even though in some cases there may still be a
degree of old-fashioned craftsmanship involved). In many ways,
the degree of sophistication achieved is only obtained through
constant feedback from users, and a better understanding of what
really happens to the foot during running. It is this combined
experience and know-how, as well as the use of advanced mate-

rials science, that eventually lead to the final design and manufacture of the sports shoe.

The future is likely to witness even better cushioning and control systems, automatic (electronically activated) lacing, and improved perspiration control. Ultimately, the purchase of sports shoes may become a completely tailor-made procedure; as in the ordering of a new car various options will be available to cope with each customer's needs, his or her whims, and (of course) size of purse.

Bicycles

4

SYNOPSIS

The bicycle, based on the 5000 year old invention of the wheel, was first tried by man about a century ago. Since then it has tended to be a forerunner in the application of innovative technology, and many inventions now associated with the development of automobiles would have been unthinkable had they not first been tried out on bikes. As with other sports equipment, today's competition bicycles are made of very advanced composite materials; on the other hand, for everyday production bikes, steel frames and wheels are proving stubbornly resilient to change.

INTRODUCTION

One of the earliest forms of bicycle to be invented was the two-wheeled Draisienne (Figure 4.1) by Baron von Drais de Sauvbrun of Baden-Württemberg. The Draisienne was first used as a walking device (Figure 4.2), to be propelled by the feet pushing directly on the road. Though these early bicycles were somewhat crude and uncomfortable (A French velocipede (Figure 4.3) was

Figure 4.1 A Draisienne designed by Karl von Drais of Paris, from 1817. From *Bicycling Science* by F. R. Whitt and D. G. Gordon (MIT Press, 1982).

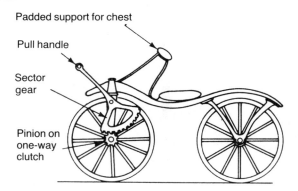

Figure 4.2 A modified Draisienne (1821) in which Louis Gompertz fitted a swinging-arc ratchet drive to the front wheel. From *Scientific American* (July 1973).

aptly named 'the bone-shaker'), production of this radical new type of vehicle propagated rapidly as its popularity blossomed. The year 1870 saw the penny-farthing bicycle (Figure 4.4) designed by James Starley and William Hillman, and the first

Figure 4.3 A velocipede designed by Pierre and Ernest Michaux in 1863 in Paris. It was the first commercial success in bicycle production, although it won a reputation at the time of being a 'bone-shaker'! From *Scientific American* (July 1973).

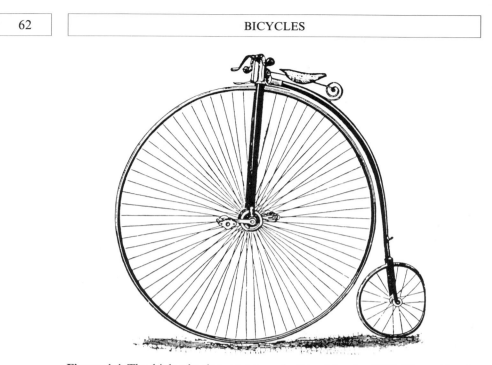

Figure 4.4 The high-wheeler, or penny-farthing bicycle, 1870. These bicycles could go faster and avoided the use of chains. From *Bicycling Science* by F. R. Whitt and D. G. Wilson (MIT Press, 1982).

'modern bicycle' concept followed in 1885 in the form of the Rover Safety Bicycle, also designed by Starley.

An important milestone in the technical evolution of the bicycle was the development by James Starley in 1870 of the tangent-tension spoked wheel (Figure 4.5), a construction which

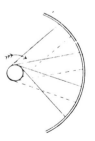

Figure 4.5 The lever tangent-tensioning spokes, invented by James Starley 1870. These spokes could be manually adjusted for optimum tension and torque, and this method remains in use today. From *Bicycling Science* by F. R. Whitt and D. G. Wilson (MIT Press, 1982).

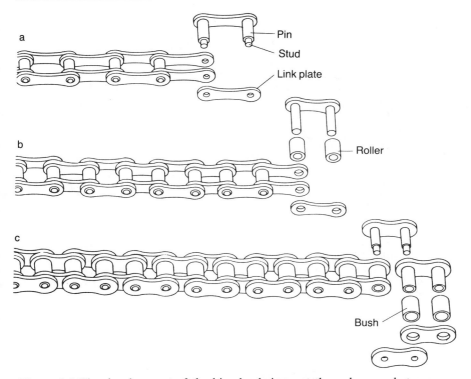

Figure 4.6 The development of the bicycle chain went through several stages before arriving at the definitive form, the bush-roller chain invented by Hans Renold in 1880. The progenitor was the pin chain, or stud chain (a), in which the pins bore directly on the sprocket teeth and the link plates swivelled on the studs at each end of the pins. In such a chain there is undue wear and friction both at the teeth and at the holes in the plates. A subsequent improvement was the bowl chain, or roller chain (b), in which friction and wear of the sprocket teeth was reduced by rollers on the pins, but wear of the plates on the studs was still too great. Renold's invention of the bush-roller chain (c), by the addition of hollow bushes to spread the load over the entire length of the pin, overcame this final shortcoming. Courtesy of *Scientific American* (July 1973), from article 'Bicycle technology' by S. S. Wilson, p.81.

could better withstand torsional (twisting) forces when pedalling on the hub directly. Another milestone was achieved in 1877 when the tubular frame and ball bearings were first adapted to existing bicycle constructions, and this development immediately gave these techniques a much wider use. Comfort and shock absorption were soon improved when pneumatic tyres were intro-

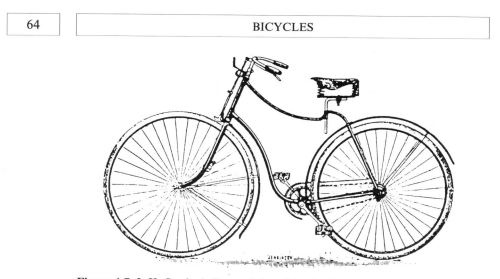

Figure 4.7 J. K. Starley's Rover Safety Bicycle, introduced in England in 1885, is regarded as the forerunner of the modern bicycle. This remarkable bicycle features rear wheel chain and sprocket drive, ball bearings in the wheel hubs, lightweight tubular steel 'diamond' frame. From *Bicycling Science* by F. R. Whitt and D. G. Gordon (MIT Press, 1982).

Figure 4.8 Modern mountain bike, with titanium alloy frame and Kevlar Bead tyres. Source: Marin brochure.

duced, this invention being first patented in 1888 by Mr T. B. Dunlop, a Scottish veterinarian living in Belfast. The bicycle chain went through many stages prior to attaining its modern form, a bush-roller chain first invented by Hans Renold in 1880 (Figure 4.6). Many of these inventions associated with bicycles were later adopted to early automobile design (Figure 4.7) and even to aircraft! Indeed, it is certainly no coincidence that Henry Ford used bicycle wheels and chains when constructing his first car, and that the Wright brothers actually began their careers as bicycle makers.

Recently bicycles have undergone a new burst of popularity with the development of the so-called all-terrain or 'mountain' bicycles (Figure 4.8). Materials requirements for these new machines are obviously more demanding in some respects because of the higher load-cycle fatigue problems when riding over rough ground or mountain tracks. Meanwhile conventional steel bikes are going as strongly as ever, and still provide the lion's share of the market.

CYCLE MECHANICS

Cycling is actually a highly efficient method of locomotion. The energy consumption per kilometre of a cyclist is in general even lower than a walker, and certainly very much lower than powered vehicles. Furthermore, not only is the sheer efficiency of cycling virtually in a league of its own, it is also achieved at an astonishing speed (quite comparable with the galloping of horses).

Riding a bike is for the cyclist a case of using the right muscles, at the right speed and in a comfortable position. Unlike walking or running, pedalling only requires muscle activity when it is needed to contribute to help boost locomotion. Muscles of course do consume energy even when the are in tension, but in this case they do not have to work continuously and at full capacity. The cyclist also preserves energy by being able to sit down; leg muscles are only put into action when they are actually needed to push on the pedals. This in itself is a beautifully natural movement in which constant rotation is achieved merely by the pedals being effectively raised and lowered. No wonder that bicycles are sometimes described in terms of being 'an extension of man', like the wearing of seven-league boots of old.

The power that the cyclist needs to push the vehicle forward depends on a number of different parameters. When pedalling on a good smooth road, a rider experiences a nice ease of rolling when pushing on the pedals. If, however, the road gets rough, or the tyres lose pressure, the cyclist soon requires considerably more force in order to maintain a constant speed. In other words the cyclist experiences a kind of 'rolling resistance'. Apart from the weight of the cyclist and the machine, the rolling resistance is also affected by the tyre pattern and its deformation or 'give', as a result of this weight. In the other extreme, train wheels have a very low rolling resistance in that they do not possess any give and the surface of the rails is quite smooth.

As a general rule, however, the largest problem for a cyclist to contend with is wind resistance. A man sitting in an upright position on a bike is far from the aerodynamic ideal. It is this aspect of course that professionals constantly work to improve in competition cycling. Greg le Mond won the 1989 Tour de France (3420 km) with just 8 seconds to spare to the runner up. His success was traced down to a clip-on extension of his handlebars, a padded tube which improved his aerodynamic shape without compromising his riding position.

Another very important parameter affecting the efficient motion of a bike obviously concerns the role played by materials used in its construction. The role of materials is important in two main respects: to ensure strength, rigidity and toughness to the frame and integral parts such as the chain and pedals, and to achieve all this without sacrificing weight saving. These aspects will now be taken up.

FROM BAMBOO TO FIBRE COMPOSITES

The average bicycle consists of over a thousand individual parts! Even admitting that half of these are located in the chain, it still leaves quite a number for the rest of the machine. In normal bicycle manufacture it is usual for many of these parts to be manufactured by subcontractors, so that bike production requires a fairly high degree of production organization. Easily the largest bicycle producers in the world are the Chinese who have an annual output of over 2 million machines, the majority of which are identical, black, and very strongly constructed from steel. They also possess the rather whimsical name of Flying Pigeon engraved on the frame.

Space frame

The frame material in early bicycles was wood, usually reinforced with metal. Indeed, wooden bikes with bamboo frames were still in use up until the end of the 19th century. Wooden models were eventually superseded by tubular steel constructions providing, as with the round cross-section of bamboo, a rigid, lightweight frame. More recently, frames have been made from rather more exotic materials such as aluminium or titanium alloys, and (in the case of competition bicycles) carbon fibre composites.

Thin-walled tubular frames are highly efficient structures designed to resist the many stresses of tension, compression, bending and torsion of cycling. In principle, the ideal type of structure for proper distribution of these various forces would be in the form of a 'space frame' as found in bridges, tower cranes, etc. Since such a construction does not really suit a bike, a modification of this had to be found and this came to be known as the 'diamond frame' (Figure 4.9). In the diamond frame bending stresses (Figure 4.10) are confined mainly to the front forks, while torsional stresses are accommodated by the frame as a whole. An alternative structure to the diamond frame is the 'cross frame' which consists of a main tube extending from the steering head directly to the rear axle and crossed by a second tube from the saddle to the bottom bracket bearings. This construction relies entirely on the strength and stiffness of the main tube. Using a 'stressed-skin' method the cross-sectional area of the main tube can be reduced accordingly, and this is the basis of today's modern cross-frame design (Figure 4.11).

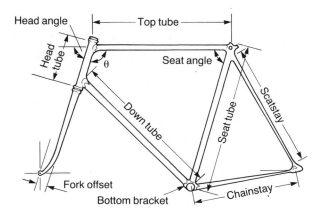

Figure 4.9 The diamond frame, typically made of chrome–molybdenum–manganese steel tubes with cast iron lugs.

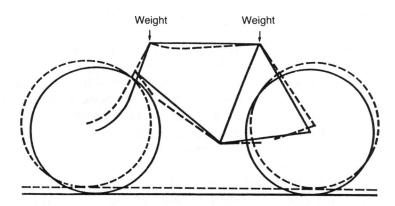

Figure 4.10 Distortions under loading of the bicycle frame.

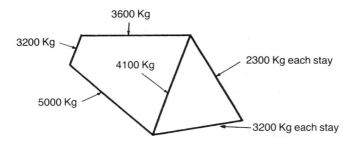

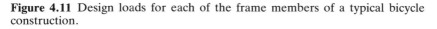

Figure 4.11 Design loads for each of the frame members of a typical bicycle construction.

Materials

The materials used for bicycle frames are dominated by steel and steel alloys such as low or medium carbon steels for production roadsters, or chromium–molybdenum–manganese steels for competition cycles. Other materials used in frames include aluminium, magnesium, titanium and carbon fibre/epoxy composites. the properties of these various materials are compared in Table 2.4.

The traditional steel-tube construction consists of several segments which are butted and brazed into the steel sockets to form the complete frame (Figure 4.12). This space-frame structure is designed to carry 100 times its own weight, a really high safety factor rarely achieved, for example, in bridge or crane construction. In military aircraft production, the factor of safety drops below two! The main problem in bicycle construction that might give rise to frame failure is a phenomenon known as fatigue, which is the result of the continuous cyclic forces of road vibration and, of course, the process of pedalling itself. The resulting steel-frame bicycle has a mass of some 2–3 kilograms. This compares with 1–1.5 kg for a carbon fibre polymer bicycle.

Recent frame material developments include a one-piece frame of a die-cast magnesium alloy (Figure 4.13) which provides extra lightness and stiffness. Magnesium is the lightest of the commercial metals used in bicycle construction, possessing only a third of the density (mass per unit volume) of aluminium. The frame is produced by using a hot-chamber, high-pressure injection-

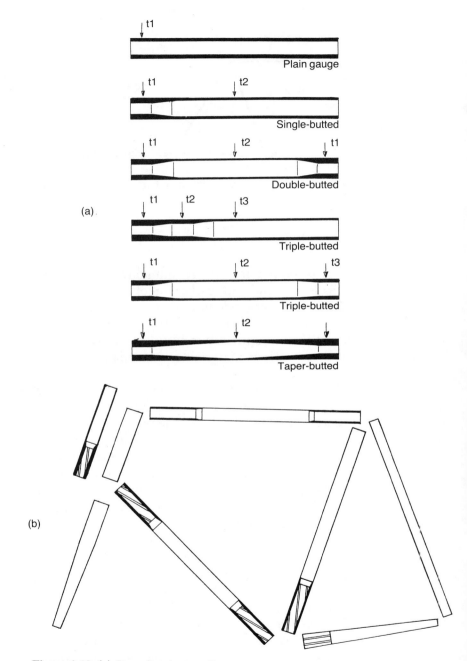

Figure 4.12 (a) Some butting profiles commonly found in main frame tubes. (b) How different profiles may be used. Source: Reynolds brochure.

Figure 4.13 A die-cast magnesium alloy frame. Note that although magnesium is only a fifth of the weight of steel and a fifth of its strength, frames need to be designed to allow for high-stress regions and this is reflected in the profile shown. Magnesium is an excellent die-casting material and machines well, too. Large illustration from *New Scientist* (30 June 1990, p.4) from 'The bike built to win' by R. Gould; inset from *Materials Engineering* (November 1989, p.33) from 'Materials at play'.

moulding technique, giving a specific stiffness increase of 50% compared with a good-quality steel bike. This material is very expensive compared with steel, however, and it is unlikely that magnesium frames will be used for ordinary roadsters.

Another material which provides a good specific strength or strength-to-weight ratio (see Table 2.3) is titanium and this metal also possesses excellent corrosion resistance. Titanium frames, while very expensive compared with steel, can be found in certain mountain bikes in order to meet their special demands in terms of high fatigue strength, toughness and corrosion resistance. Only top-of-the-range mountain bikes and competition racers use titanium today.

The British bicycle manufacture, Raleigh, has recently produced a frame based on a metal matrix containing ceramic fibres. This is an advanced materials concept in which an alloy of copper and aluminium provides the 'binder' for silicon carbide ceramic fibres. While extremely expensive, these frames are practically as light and strong as carbon fibre polymer frames, and have certain advantages in overall construction of the bicycle. For example, the problems of joining one component to another should be less for the metal matrix composites compared with the polymer-based ones.

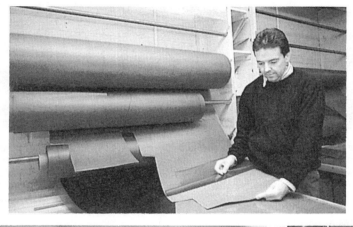

Figure 4.14 Carbon- and Kevlar-fibre-reinforced epoxy bicycle frames are made up from sheets of the composite shown in (a). After profiling, the whole is baked and cured in an autoclave to produce the finished item shown in (b). Note the special bolted joints for attaching other components to the frame, since these composites cannot be welded. From *Cycling Weekly* (2 March 1991, p.6).

Bicycle frame construction is a very challenging area for application of the new lightweight fibre composite materials. These new materials provide the frame with a lightweight construction possessing properties of very high strength and toughness, high stiffness and excellent fatigue and corrosion resistance. However, like magnesium and titanium, these materials are extremely expensive and only the cycling elite can presently afford them. For this reason, tubular frames made from composite materials are still only available from a few manufacturers. Most of these advanced fibre composite constructions rely on the special features of carbon, Kevlar and ceramic fibres. These include high stiffness (the carbon fibres), good toughness (the Kevlar fibres) and adequate damping resistance (the ceramic fibres). A typical construction used in frames is a carbon fibre tubing, with cross-woven carbon fibres impregnated in a polyester or epoxy resin (Figure 4.14). A layer of Kevlar is added to protect the delicate carbon fibres (Figure 4.15) from scratches and to improve tough-

Figure 4.15 This illustrates another solution (besides bolting) to the joining problem, i.e. carbon fibre composites. In this case the hydraulic brake is attached to the composite forks by a special polymer glue. From *Cycling Weekly* (2 March 1991, p.7).

Figure 4.16 The first all-polymer bicycle. The unconventional design exploits the good formability of polymer composites.

ness. Additional reinforcements at the bottom bracket, seat and head are usually added to increase the frame's stiffness even further. The lugs usually consist of a suitable light metal alloy, e.g. aluminium or titanium, while the rear reinforcing bridge is normally made of a glass fibre composite.

All-polymer bikes for everyday use have now been manufactured. They have wheels made of glass-fibre-reinforced polyamide, while the frame is of glass-fibre-reinforced polyester, produced by an injection-moulding technique. These bicycles (Figure 4.16), however, have not proved a commercial success in that they are rather cumbersome in looks and lack the sleek form of a tubular steel bicycle.

The wheels

It may seem useless to try to improve such a well-tried component as a bicycle wheel. Off-road bikes, however, require more stable and rigid wheels than a normal bike, and this is one area where glass-fibre-reinforced nylon has been tried out in wheels (Figure 4.17). Another application where there is room for improvement is in competition cycles where aerodynamics and lightness are of

Figure 4.17 Kevin Gill in action during a time trial, with a wheel combination typical on the racing scene today. He uses a tri-spoke-style front wheel and a disc wheel on the rear. There is a definite speed gain – if you can meet the cost. From *Cycling Weekly* (2 March 1991, p.20).

Figure 4.18 Two commercial examples of composite spokes wheels. These wheels have an advantage over disc wheels in cross-wind conditions and as shown in Figure 4.17 are particularly suited to the front wheel. A recent study by Dupont has shown that the drag coefficient in the sleek aerodynamic three-spoke wheel is reduced to only 0.06 compared with a conventional wire-spoked wheel which has a coefficient of 1.2. Both the above wheels are made of carbon/Kevlar-fibre-reinforced epoxy and possess aluminium rims and steel axles. From *Cycling Weekly* (2 March 1991, p.21).

great importance. This has led to the disc wheel. These discs are usually made of an aluminium alloy or a carbon fibre composite which replaces the spokes in the wheels. The explanation of this is to do with the conservation of momentum that is achieved; this, combined with an improved rigidity, is a feature that is essential in pursuit races for example. The latest development combining lightness, rigidity and improved cross-wind aerodynamics is the three- or five-spoked wheel (Figure 4.18)

Optional extras

Frame and wheels are not the only part of a bicycle which take advantage of new materials. Bicycle seats, for example, are now moulded from lightweight nylon (Figure 4.19), giving a weight reduction of up to 300 g or half a pound. Brake levers, gear changes and power switches are all moulded from acetal resin;

Figure 4.19 Saddles provide many problems to cyclists and modern saddle design needs to consider a range of parameters. The example is a test saddle at a laboratory in The Netherlands wired to give information on pressure points. From *Delft Outlook* (published by Delft University, February 1991, p.18).

this resin is stiff and resists impact as well as being lighter in weight than metallic materials. Recent research on the development of new pedals is based on a kind of mini-ski binding and is made of a steel–magnesium alloy, allowing the rider to unclip his or her feet more easily. A sideways flick releases the foot from the pedals just as in downhill skis and these fixtures are thought to be safer than conventional leather straps. Other systems like the 'Time' pedal allow the foot to swivel a few degrees, and thus the risk of injury to knee tendons can be reduced. The simplest safety pedal, however, connects the rider's shoe by a crescent-shaped clip to the pedal. This 'Aerolite' system uses bearings similar to those found in the wing components of jet aircraft.

FUTURE TRENDS

Steel-frame bicycles have proved so popular over the last century that it would perhaps be folly to suggest that conventional bicycles will ever be taken over by new materials and composites, particularly considering the high cost involved. On the competition front, on the other hand, decreases in weight will continue to occupy the efforts of cycle manufacturers. For example, the Japanese have recently announced the first all-paper bicycle! The frame of this bicycle is constructed from hand-laid-up paper and epoxy resin. The resulting cellulose fibre alignment provides a strength which is as much as 60% of that of carbon fibre composites, no mean feat indeed. The resulting frame has a mass of only 1.3 kg, compared with 1.2 kg for the best carbon fibre frame. A thin plastic coating encases the paper to ensure that the bike does not collapse into a soggy heap in the rain!

Tennis and squash rackets

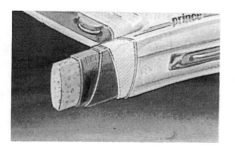

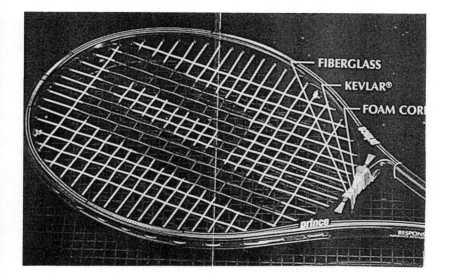

5

SYNOPSIS

The racket sports are the fastest expanding of all leisure activities. Any participant is these sports knows, too, that there is nowadays an enormous range of rackets available, and that high-tech descriptions such as graphite fibre or ceramic fibre composite abound. In this chapter the pros and cons of these descriptions are analysed and discussed. In common with other advanced sports equipment, however, it will be found that the emphasis of racket design and construction is on optimum performance, combined with freedom from injury. The chapter concludes with details of the design and construction of tennis and squash balls; even these items have not escaped technology's impact.

INTRODUCTION

Traditionally, rackets have until quite recently been made of ash, maple and okume. Laminated constructions of these woods were then developed to cope better with the high impact forces encountered in tennis and squash. In the late 1960s, however, metal frames, usually steel or aluminium, were introduced. From research in biomechanics, it was discovered about this time that the early rackets were poorly constructed to damp the high vibrational forces which are generally regarded as the main cause of 'tennis elbow' (Figure 5.1). Today's composite constructions thus seek to combine several desirable properties with the objective of improving the racket's strength and durability, as well as damp the high impact forces involved in these sports. We shall first consider the biomechanical background of racket design and then discuss how the new advanced materials are used to improve performance and even help protect the player from injury. We shall also look at the design and manufacture of tennis and squash balls.

BIOMECHANICAL CONSIDERATIONS

Seen from a biomechanical point of view, rackets, like any other striking implement, are in effect used to provide the striker with

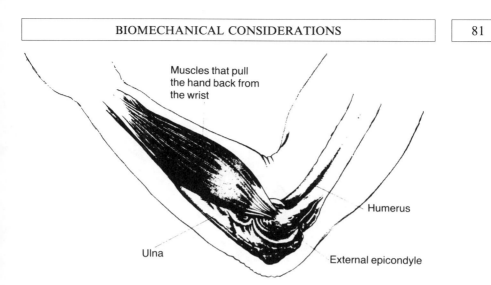

Figure 5.1 'Tennis elbow'. Symptoms and diagnosis:

- Pain which mainly affects the outer aspect of the elbow, but can also radiate upwards along the upper arm and downwards along the outside of the forearm.
- Weakness in the wrist. This can cause difficulty in carrying out such simple movements as lifting a plate or a coffee cup, opening a car door, wringing a dishcloth and shaking hands.
- Pain over the lateral epicondyle when the hand is bent backwards (dorsiflexed) at the wrist against resistance. This sign alone is sufficiently important to justify a diagnosis of 'tennis elbow'.
- Pain on the outer aspect of the elbow triggered by straightening the flexed fingers against resistance.

an extra dimension, or arm, to the game. From a purely technical point of view, they may be regarded as a tool to impact and accelerate the speed of the ball efficiently. The forces involved here (Figure 5.2) thus vary according to the weight of the ball, the material of the racket, the string tension and also the power of the stroke (the human factor). Basically, all of these parameters are the same for tennis, squash or badminton.

The impact force on a player returning a tennis ball travelling at 150 km h^{-1}* is roughly equivalent to jerk-lifting a weight of about 75 kg (170 lb). It is a sobering thought, but such are the forces that have to be accommodated by the player's frame. If a player, however, uses a faulty technique instead of hitting the

*These speeds are typical of those achieved by top players.

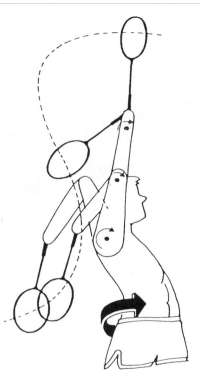

Figure 5.2 The biomechanics of hitting a tennis ball. This action can be compared with throwing a ball in that initially the arm is bent back, and the muscle groups are activated in the order: abdomen, shoulder, elbow, wrist. From *Athletic Ability and the Anatomy of Motion* by R. Wirhead (Wolfe Medical Publications, 1980, p.105).

strokes with a firm wrist and a well-balanced stance, the vibration set up in the player's arm on impact may be hard to accommodate without causing injury.

The forces normally transferred to the bone structure are distributed to the main muscles of the arm causing a high load to the lateral epicondyle, located on the outer side of the elbow. These forces may in turn damage the small blood capillaries in the muscles and tendons around the elbow joint.

A recent study has shown that 45% of all players who play tennis regularly suffer from these problems! Indeed, apart from the player improving his or her technique, the only way of combating this problem lies in improving the design of the racket itself. This in turn is a function of the design and clever application of appropriate materials combinations.

RACKET MECHANICS: THE SWEET SPOT

The philosophy behind modern racket designs is to improve the dynamic characteristics, to make ball control easier, and to damp dangerous vibrations set up by the stroke itself.

It is generally assumed that the ball is hit 'somewhere in the middle' of the racket, the so-called 'sweet spot'.

The sweet spot is defined as:

- that part of the strings at which the player feels no shock following impact;
- that part of the strings that has a minimum of vibration when impacted (this is also referred to in vibration mechanics as the 'nodal region').

Using advanced techniques (Figure 5.3) such as multi-dimensional spectral analysis and nodal analysis, it has been shown (Figure 5.4) that the damping characteristics of the racket, the attachment of damping materials to the grip, and changes in stiffness all affect the size of the sweet spot. In other words, the size of the sweet spot is likely to vary greatly from racket to racket, and that the type of materials used in the racket construction have a direct effect on its shape and size.

MAKING A RACKET

The head

It has been shown that the overall performance of a racket ultimately depends upon the skill of the player in combination with the design and material of the racket; however, research shows that the shape of the head is also an important factor to be considered. In modifying the head shape it is also possible to enlarge or to move the power and nodal regions, and hence enlarge the sweet spot. Rather appropriately the Head Corpora-

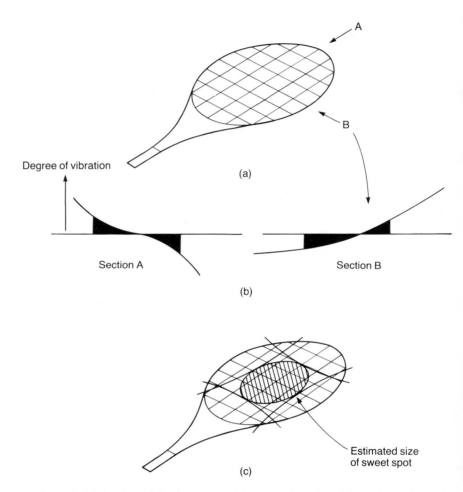

Figure 5.3 The size of the 'sweet spot' in a tennis racket (a) can be estimated by measuring the various vibrational modes acting across the racket after impact with the ball. A ball impacted within the sweet spot sets up very little vibration, as seen in the measurements of (b). By summarizing these vibrations in different cross-sections of the racket, the size of the low-vibration sweet spot can be estimated as shown in (c). After J.-E. Oh, H. Park, Y.-Y. Lee and S.-H. Yum (1987) Estimation of nodal parameters due to the change of sweet spot by structural modification of a tennis racket, *JSME International Journal*, **30**, 1121.

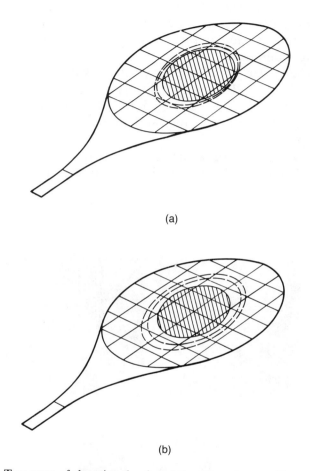

(a)

(b)

Figure 5.4 Two ways of changing the size of the sweet spot are illustrated here. In (a) the size of the sweet spot has been increased by increasing the stiffness of the racket, for example by changing from an aluminium frame to a carbon-fibre-reinforced polymer frame. In (b) the sweet spot has been increased by changing the size (mass) of the handle and frame, keeping the material the same. After J.-E. Oh, H. Park, Y.-Y. Lee and S.-H. Yam (1987) Estimation of nodal parameters due to the change of sweet spot by structural modification of a tennis racket, *JSME International Journal*, **30**, 1121.

tion (USA) made the first range of oval-shaped heads in 1976, and in doing so the impacting area was immediately enlarged without changing the head's length. Since then manufacturers have constantly experimented with sizes and shapes other than oval in order to make further improvements to the racket's impacting qualities.

Today egg- or drop-shaped designs, with slight variations (Figure 5.5), are all to be found in sports stores. The design of the head, however, is restricted by the International Tennis Federation which has now imposed an upper limit for the size of the racket.

The development of squash rackets is to some extent similar to that of tennis rackets. Up to 1983, frame construction was

Figure 5.5 Illustrating different shapes of racket heads: (a) tennis; (b) squash. Courtesy of Wilson.

always of wood, a material often criticized by players because splinters would fly if the racket was accidentally smashed against the wall of the court. Today's constructions are thus mostly made from metal alloys (mainly aluminium) or fibre-reinforced plastic (FRP). The alloy construction is made from a length of hollow extrusion which is lasted together at the end facing the grip. Though there have been several attempts to change the head shape of squash rackets (as in tennis rackets) most of them still retain their traditional circular form. The use of FRP materials (carbon or glass fibre combinations) gives the advantages of lightness, strength, toughness, less vibration and greater freedom from splintering when fractured.

MATERIALS IN RACKET CONSTRUCTION

Aluminium, formed in a tubular streamlined profile, was the first material to follow the traditional wooden constructions (solid or laminated). Tests revealed, however, that these aluminium rackets exhibited plastic (irreversible) deformation and a notable loss of rigidity after around 6000 impacts. It soon became clear that the multitude of properties needed in tennis, squash or badminton could not be fulfilled by a single material. As a result of this, today's racket constructions are based on composite materials consisting of multiple fibre-reinforced layers wrapped around a soft inner core. This core is mostly an injected polyurethane foam or even a honeycomb structure (Figure 5.6). The core material is important in reducing the racket's high-frequency vibrations when impacted.

Apart from these shock-absorbing and dampening features, a racket also needs stiffness (to impart high forces to the ball) and resilience. That is where high-performance fibre reinforcements such as carbon, aramid or E-glass come in. These advanced materials increase the racket's life from (typically) 5000 impacts for a wooden/ox-gut construction, to over 30000 impacts for a modern FRP racket containing nylon strings. It has been shown that rackets of carbon-fibre-reinforced polymer lose a mere 3–4% of their stiffness during a lifespan of up to 50000 impacts.

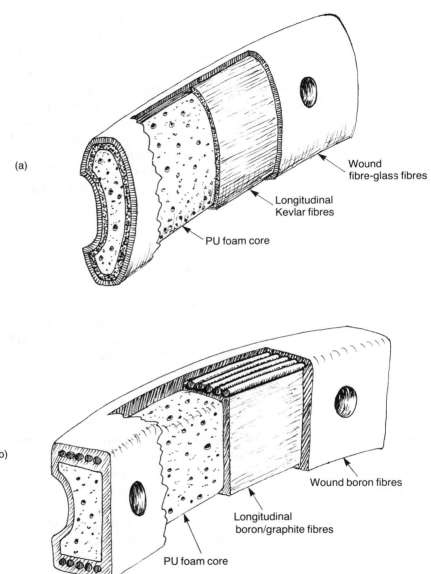

Figure 5.6 Two possible solutions to obtain high-strength, high-stiffness racket formes. In (a) the slightly cheaper glass fibre outer wound fibres are used, which cover the high-stiffness and high-dampening Kevlar fibres. In (b) the high-strength boron fibres are wound around longitudinal boron/graphite fibres, the latter combination providing both very high strength and stiffness. Racket (a) is kinder to persons with problems of tennis elbow; racket (b) is for the highly professional.

Fibre reinforcements are usually made from unidirectional fabrics, monofibres or hybrid braids.

SPECIFIC DESIGNS AND TESTS ON RACKETS

One or two examples may best illustrate how manufacturers design their rackets with the help of tests in which various combinations of materials are laid up in slightly different ways in order to achieve optimum properties. The Wilson Ultra FPK racket consists, for example, of a urethane core, 84% graphite, 12% Kevlar and 4% Fiber FP (weight per cent). The graphite provides strength and stiffness which minimizes head deflection, and helps prevent twist of the racket head when the ball impacts outside the sweet spot. Kevlar fibres provide additional strength and durability but also damp the racket's vibration. Fiber FP, a new ceramic fibre (a pure form of aluminium oxide), provides even greater stiffness and damping to this range of racket designs.

Optimum racket design and construction is being assisted in carefully monitored mechanical tests in which a hammer strikes the racket at various points, setting up resonance frequencies which affect the racket's absorption and other features. These results tend to indicate that the glass/carbon (80/20) racket is the least rigid, while the aramid/carbon (10/90) is the stiffest of all. The distribution of nodes in these tests indicates the smallest deformation that a racket is likely to encounter during play, which in turn is also an indicator of any overloading likely to be subjected to the arm muscles; in other words, this test is claimed to show when fatigue (tennis elbow) may occur, and provides an indicator of how to optimize racket design in this respect.

In practice, of course, there are several design/material combinations to choose from, and every manufacturer claims their own 'unique' designs and properties. However, it cannot be disputed that new advanced composite materials do contribute significantly to performance, and that they are likely to help achieve freedom from injury particularly to leisure-time players.

RACKET STRINGS

Traditionally, the strings used in rackets were ox-gut. It is now estimated, however, that the demand for tennis rackets is so great today that there simply would not be enough oxen to go round! As a result of this, various types of synthetic (e.g. nylon) strings are used (Figure 5.7).

Apart from the material that the strings are made of, the string tension is vitally important to the racket's performance. This string tension interacts in a fairly complex way with the racket head and ball deformation during impact. Clearly, the higher the string tension, the greater the deformation of both head and ball. A recent study has shown that ball rebound velocity was lowered by about 6% when the racket was strung at 170 N, compared with 220 N. In addition, it was found that rebound velocities increased when nylon stringing was used rather than ox-gut (because of the higher Young's modulus of nylon). The racket's flexibility is another parameter that interacts with string tension.

Figure 5.7 Anatomy of a single racket string. Each 'string' is seen to consist of entwined bundles of very fine, high-strength nylon fibres. Individual bundle strands are immersed in a matrix of a polymer elastomer solution and the whole string is protected from moisture and wear by a polymer coating. Courtesy of Dunlop.

Research suggests that medium- and high-flexibility rackets should be strung at a lower tension (245 N) to achieve higher rebound velocities, whereas low-flexibility rackets ought to be strung at higher tensions (290–335 N). Another recent innovation brought out by a few manufacturers is to fill the strings with oil. In this process the strings are produced with a fine central core for accommodating the oil. It is claimed that such strings improve resilience and reduce internal friction, thus increasing the string's lifecycle time.

While all this is useful for racket designers and manufacturers, the leisure players will themselves inevitably find their own ideal racket types and corresponding string tensions. In most cases, of course, the question of racket cost has also to be balanced against quality and brand. Indeed, kilogram for kilogram, it has been estimated that a good high-tech racket costs rather more than a jumbo jet!

BALL CONSTRUCTION

Ball sports have been played since ancient times. Apparently the useful elastic property of natural rubber to deform and then snap back to its original size has intrigued people for centuries. This elasticity is, of course, the principal feature that is needed for tennis, squash and even golf balls. In practice, the ball needs a number of additional properties besides good elasticity, notably resilience and wear.

On impact, energy is stored within the ball, and this then leads to the ball's ability to rebound. While the detailed mechanism of the impact between ball and bat or club is hard to model, any player knows and feels whether the click of a golf ball on contact with the clubhead, or the 'response' of the ball on the sweet spot of the racket strings, is 'right' or not. As we shall now show, ball construction is a sophisticated technology today; it is also an extremely profitable one. In 1980 alone, it was estimated that 300 million golf balls, 180 million tennis balls and 6 million squash balls were produced. These numbers have grown considerably today.

Tennis balls

Tennis balls were originally made from a solid rubber core covered by a flannel stitching. This was later improve by making the core hollow and pressurizing it with gas. Chemicals generating gas within the ball are added prior to the balls being sealed. In order to achieve a uniform wall thickness and a high degree of reproducibility, the separate twin half-shells need to be precision compression moulded. The flannel cloth is now replaced by a synthetic nylon and wool composite polymer, and the cloth stitching used earlier is replaced by a vulcanized rubber seam. In order to inflate the core of tennis balls, two methods are used: either chemicals can be employed, usually sodium nitride and ammonium chloride, which produce nitrogen gas; alternatively, air can be compressed into the ball using a complex clam-shell press arrangement. Core materials for pressurized balls can be based on natural rubber containing a high density of fine-particle filler material, possessing properties of low gas permeability. After a shelf life of only 4 months, however, balls made by these techniques were found to be unable to meet the required pressure standards, even when stored in special sealed cans. To overcome this problem, pressureless balls have now been developed from a highly resilient, high-modulus polymer compound. The various polymer materials used in balls include high-styrene resin, celluloric fillers, aramid plastic resin, and (more recently) unsaturated carboxylic acid copolymers (combinations of two or more polymers). The outer casing cloth contains a high wool content (50–60%) but for optimum bounce properties, the balance of the fibre content is made up by nylon fibres.

In former times it was the skill of the ball craftsman that produced the perfectly spherical surface of the finished article, which decided the flight and rebound qualities of the tennis ball. Today, however, balls are produced in seconds using fast injection-moulding techniques, in conjunction with fully automated, computer-controlled machine tools.

Squash balls

In contrast to tennis, squash players have a choice of four

different types of ball, each of which varies in size, weight and rebound properties.

Apart from these variations, the dynamic behaviour of a squash ball is also affected by temperature and the material of the court wall. In addition, a suitable value of surface friction is required to create just the right range of rebound angles. The high deformation of an impacted squash ball demands in particular good durability with respect to the joint between the two half-shells of the ball.

Traditionally, squash balls have been made from carbon-black-impregnated rubber compounds which tended to mark the walls of the squash court. However, most balls are now based on 'non-marking' polymer materials. The manufacturing process of a squash ball is not unlike that of a tennis ball; the exact material specification is, however, a closely guarded secret!

6

Boats and boards

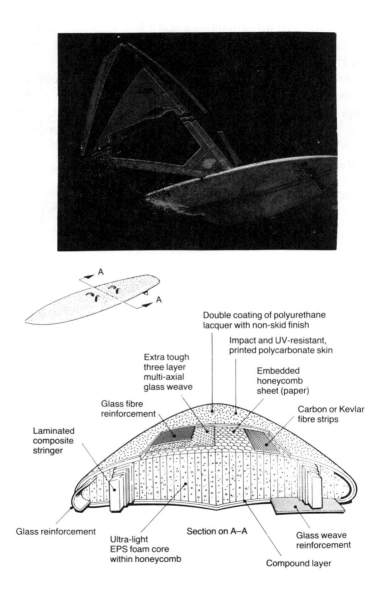

Double coating of polyurethane
lacquer with non-skid finish

Impact and UV-resistant,
printed polycarbonate skin

Extra tough
three layer
multi-axial
glass weave

Embedded
honeycomb
sheet (paper)

Glass fibre
reinforcement

Carbon or Kevlar
fibre strips

Laminated
composite
stringer

Glass reinforcement

Ultra-light
EPS foam core
within honeycomb

Section on A–A

Glass weave
reinforcement

Compound layer

SYNOPSIS

Traditionally, boats and surfboards have been lovingly constructed from wood. From an aesthetic point of view no one can deny that wooden boats in particular look, feel and even sound very much nicer than the oven-baked plastic jobs. Nevertheless, plastic boards and boats, based mainly on glass fibre and other fibre-reinforced epoxys, are generally a good deal cheaper, easier to maintain and in many ways safer than those made of wood. Both boards and boats are beginning to use surprisingly advanced materials concepts in their basic construction.

INTRODUCTION

While glass-fibre-reinforced epoxy boards and boats are definitely not as aesthetically attractive as reed rafts, canvas canoes or wooden dinghies, they are generally lighter, more portable, require less maintenance and are cheaper. At competition level, advanced (non-wooden) materials nowadays dominate board and boat construction for reasons of lightness, lower skin friction, higher toughness and better safety.

Irrespective of the type of watercraft concerned, there are four basic forces to contend with when considering boat design (Figure 6.1). These are: weight, lift, thrust and drag. Weight is simply the gravitational force acting on the craft due to the mass of boat and passenger. Lift is generated by the boat's buoyancy, i.e. the displacement of water by the craft's hull. On the other hand, when the boat moves, 'dynamic lift' occurs and this in turn reduces the buoyancy force. Thrust and drag are the forces that decide the speed of the boat; thrust propels the boat along and the drag forces resist the boat's motion (Figure 6.2) due to friction and bow resistance. Obviously these resisting forces are mainly a function of the hull's design. However, certain materials with lower skin friction can be incorporated in the hull. Skin friction in this case is a fairly complex phenomenon involving a smooth (laminar) component of flow and a turbulent (chaotic) one (Figure 6.3). Since turbulent flow increases drag, skin resist-

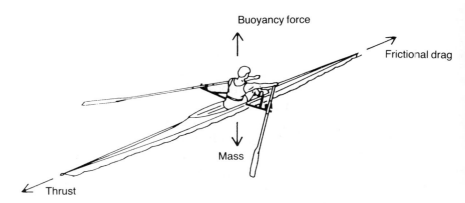

Figure 6.1 Forces acting on a boat. Modern materials can help to reduce weight and minimize frictional drag. From *Tomorrow's Materials* by K. E. Easterling (Institute of Materials, 1990).

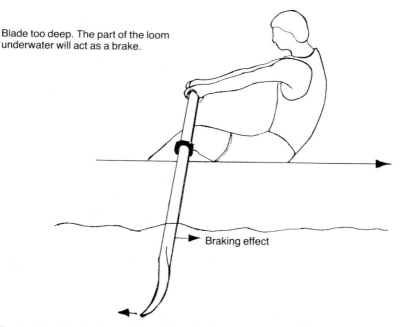

Figure 6.2 The braking effect if the blade is too deep. This effect plus water turbulence are other forces that need to be taken into consideration when estimating total drag acting on a boat. From *Rowing and Sculling* by Bill Sayer (Robert Hale, 1991).

Figure 6.3 An underwater photograph of the fin of a windsurfer, showing fairly non-turbulent streams of bubbles moving from right to left across the fin. However, the fine white line emanating from the bottom of the fin is a vortex of flow which adds to drag on the fin. The fin in this case is on an advanced composite/polymer construction and contains fine grooving at the surface to reduce onset of turbulence. From the research of Dr Francis Chui of the University of Exeter.

ance can be reduced if the transition from smooth to turbulent flow is delayed as long as possible as the boat picks up speed. NASA has studied this phenomenon and found that waxing the hull's surface has little effect on skin friction. If, however, the hull's surface contains fine grooves running in the flow direction, drag can be reduced slightly. This effect was utilized in the *Stars and Stripes*, winner of the America's Cup competition in 1988. By coating the aluminium hull with fine-grooved polymer material, the yacht's speed was increased by 2%. Although this may not sound much, it proved to be pretty decisive in the finals, providing a four-boat-length advantage when measured over the whole course.

MATERIALS FOR RACING HULLS

The key objective in boat design is to minimize drag at the normal operating speed of the boat. The obvious approach is to have a

Figure 6.4 The photograph of a four-man racing boat well illustrates the relative fragility of the lightweight hull with respect to the substantial mass of the rowers and forces from rowing. This has led to the need for advanced fibre composite construction of the hull in order that it can withstand the large forces involved. From *Rowing and Sculling* by Bill Sayer (Robert Hale, 1991).

shell that is lightweight, long and narrow. In the past, racing-boat shells were beautifully crafted from cedar, spruce or mahogany (Figure 6.4) and they were made lighter by constructing their hulls as thin as paper. These shells were prone to damage and many a careless finger penetrated the hull's shell during handling. The 1950s witnessed the development of glass fibre/polymer hulls and by the late 1960s fibre composites began challenging the dominance of wooden boats. Today, of course, wooden shells in competition sculls and yachts are a rarity indeed.

Because of the special demands of water-based sports, competition boat hulls invariably consist of a combination of several composite materials. The objective in having this degree of sophistication is to obtain a hull that is tough and strong, and is at the same time as light as possible. A lightweight single-scull boat can be as light as 10 kg. Most types of hull rely on polymer/glass fibres, with Kevlar or carbon fibres often present for extra toughness and strength. A good racing hull, for example, may typically consist of a sandwich construction based on alternate layers of glass fibre mat and Kevlar woven fabrics (Figure 6.5), bonded with a suitable core material (e.g. ethafoam). The core material is a cellular polymer and provides lightness without loss of stiffness. The hull bottom and the strengthening laminates for the hull sides are usually made from two or three double layers of glass fibre–Kevlar mixed composites.

If more local rigidity is needed, a complex composite consisting of a polyester isophatalic resin and plywood encapsulated in a Kevlar weave has been used. As a matrix material, epoxy or isophatalic resins (NPG) are nowadays beginning to be considered as standard for this type of competition vessel, a development that has come about because of the need to protect the outer hull against blister formation. The structural weakness that follows this blister phenomenon has been found to lead to leakage through glass fibre osmosis, a curious polymer deterioration due to the breaking down of the bonding agents holding the fibres and the epoxy resin matrix together. Problems of osmosis are also widely reported for competition canoes and the phenomenon appears to be aggravated by the high stresses involved in competition racing conditions.

The shells of competition single-scull boats are also nowadays made from Kevlar aramid fibre/carbon fibre composites, which ensure good rigidity and lightness. This solid laminate construction (no core material in this case) (Figure 6.6) is found to be 25% lighter, 50% stronger in tension, and exhibits a 50% improvement in hull rigidity compared with the cheaper glass fibre composite hulls.

Figure 6.5 A monocoque construction eight made from sandwich laminates of carbon and Kevlar on a honeycomb core, giving great stiffness and strength with minimum weight (Aylings E mould). From *Rowing and Sculling* by Bill Sayer (Robert Hale, 1991).

Figure 6.6 A single-scull constructed from kevlar aromid fibre/carbon fibre/epoxy composite, ensuring good rigidity and lightness. From *Rowing and Sculling* by Bill Sayer (Robert Hale, 1991).

In order to improve the performance of racing catamarans a new material design for the structural beams spanning the hulls has recently been introduced. This is a board of graphite–glass mixed fibre-reinforced epoxy resin composite, and has several advantages compared with simple glass-fibre- or aramid-fibre-reinforced epoxy resins. This is because it is difficult to meet the competing demands of both lightness and a really tough design at the same time. Hybrid materials, such as aligned graphite fibres together with angular ply-glass are made in elliptical cross-sections and these are found to have certain structural advantages. As far as weight saving is concerned in these catamarans, a combination of a pitch 55% epoxy resin composite, in the beam spans, together with 50% by volume axially orientated glass/carbon fibres in the hull construction, produces a weight saving of up to 14% compared with straight glass fibre composite material in these vessels.

Another approach being employed to improve the strength and rigidity of racing-boat hulls is to use a combined bi-directional and uni-directional carbon fibre in the form of a woven fabric, which is constructed and laid up by hand. In the case of yacht masts, these are now produced by winding the carbon fibre filament in a similar way to that discussed earlier in connection with vaulting poles. In this case several layers of fibres are arranged both longitudinally and circumferentially in order to combine flexible stiffness with strength. Optimization of the use of these expensive carbon fibre materials is achieved by controlling the density of the winding using a computer-controlled manufacturing process.

CANOES AND KAYAKS

For this group of competition boats polyethylene, polyurethane and polyester composites are often preferred instead of epoxy resins in order to gain optimum toughness. Cross-linked polyethylene has a superior impact resistance when compared with glass- or graphite-fibre-reinforced materials. Sprayed rigid polyurethane foams are also applied to the interior hulls of these craft, providing them with a high degree of buoyancy and additional rigidity. Structural polyurethane foam is also used in the production of dagger boards to provide better high-speed stability. The use of these advanced fibre composites with cellular (foam) linings undoubtedly greatly improves the toughness of canoes and kayaks, and this must be a vital consideration in a sport where the vessels have often to be navigated through dangerous and rocky streams.

OARS AND PADDLES

The traditional material for oars and paddles is, of course, based on laminated wooden strips. However, considerable skill and handwork is needed to produce these, and the finished product is correspondingly expensive and may even possess a number of weak points (Figure 6.7). For example, an ultrathin wooden blade tends to be very fragile, while the shaft needs constant attention if it is to retain its shape and stiffness. Repairs of laminated wooden oars are obviously difficult, while simple blade

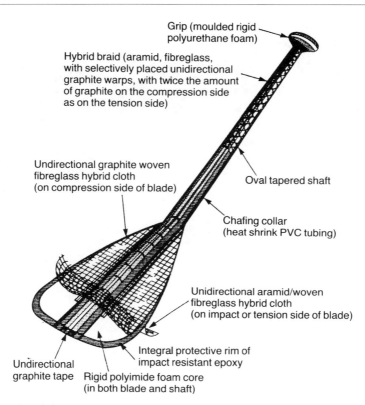

Grip (moulded rigid polyurethane foam)

Hybrid braid (aramid, fibreglass, with selectively placed unidirectional graphite warps, with twice the amount of graphite on the compression side as on the tension side)

Undirectional graphite woven fibreglass hybrid cloth (on compression side of blade)

Oval tapered shaft

Chafing collar (heat shrink PVC tubing)

Unidirectional aramid/woven fibreglass hybrid cloth (on impact or tension side of blade)

Integral protective rim of impact resistant epoxy

Undirectional graphite tape

Rigid polyimide foam core (in both blade and shaft)

Figure 6.7 A bent-shape canoe paddle is made of a hybrid composite that is high in strength and ultralight. The paddle is tough, yet provides smooth and efficient propulsion. Courtesy of Tertm Corp.

replacement is prohibitively expensive. The design and construction of a reinforced polymer oar is thus an excellent example of how advanced materials can be tailored to meet the demands of a very special application.

The long central part of the oar or paddle is usually crucial in oar construction, when considering the desired properties of good stiffness, flexibility and allowable shaft deflection. A typical force applied by a rower or paddle can be up to 1000 N giving rise to very substantial bending stresses (Figure 6.8) in the slim shafts of oars. This consideration, together with the need for lightness and adequate toughness, obviously call for a demanding performance from both man and material. The resulting oar construction

Figure 6.8 This photograph shows the start of a coxed pair race at Ghent. The bending and larger strain on the loom of the oars is apparent as maximum power is applied. This has led to the use of carbon fibre/epoxy composite materials in racing oars. Besides providing high strength and stiffness there is also a weight saving of about 1 kg or more compared with laminated wood constructions. From *Rowing and Sculling* by Bill Sayer (Robert Hale, 1991).

consists of four separate components, each part produced requiring in effect a quite different approach. It is interesting to consider in a little more detail how these component parts are produced in order to illustrate the degree of sophistication now achieved in oar construction.

The shaft

This consists of several fibre composite layers, each layer laid up in the form of filament windings. Filament winding is actually made on a special machine initially designed for the production of ship's masts and ski-poles. The individual layers are tensioned along the longitudinal axis, and are also wound circumferentially (Figure 6.9). Extra reinforcements are also made where the three other component parts are later to be attached.

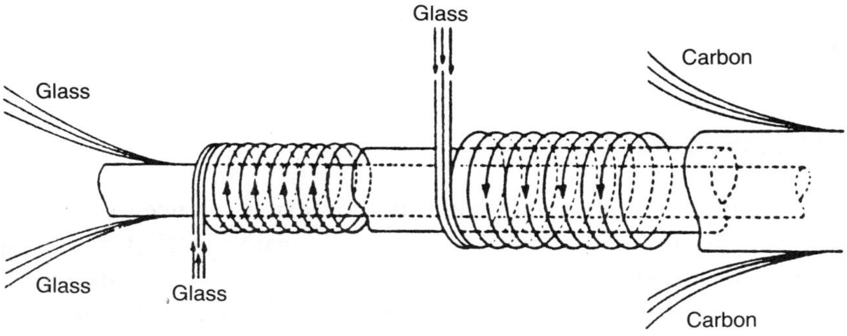

Figure 6.9 This schematic illustrates the construction of the shaft (or loom) of an oar, in which three layers of glass fibres are reinforced by an outer layer of carbon fibres, the whole being embraced in an epoxy resin. J. H. Prinsen (1986) Design and construction of a reinforced plastic oar for use in competition and recreational rowing, *Polymer Composites*, **7**, 227.

The blade

This is a particularly complex component to construct. Besides the need to be strong and tough, it requires an extremely thin cross-section for easy entry in and out of the water. This is achieved by having two external layers of fibre glass cloth, these being layed down just prior to hardening of the epoxy. The interior cavity of the blade is filled with a low-density structural cellular foam for extra lightness. To stiffen the blade further a hollow fibre tube is also inserted into the core of the blade and encapsulated within the foam core.

The sleeve

This is needed for the adjustment of the collar (the joint between the blade and the shaft). This component is made from a thermosetting polyurethane plastic which is hand poured into moulds and then appropriately clamped at the shaft.

The handle

This is turned from kiln-dried sections of Ontario poplar. The length of this wooden insert may be varied to achieve a well-balanced oar, with its centre of gravity roughly at the same point as for the wooden oar.

SURFBOARDS

Surprisingly though it may seem, the materials science of 'simple' windsurfing boards is every bit as sophisticated as that of racing boats. The board's core (Figure 6.10) consists of an extruded foam polystyrene filler enclosed in fibre glass. Wound round the core are a number of spun graphite fibre strands embedded in a PVC resin matrix. This PVC/fibre composition is enclosed in turn by four layers of high-stiffness E-glass fibre weave. Finally, the

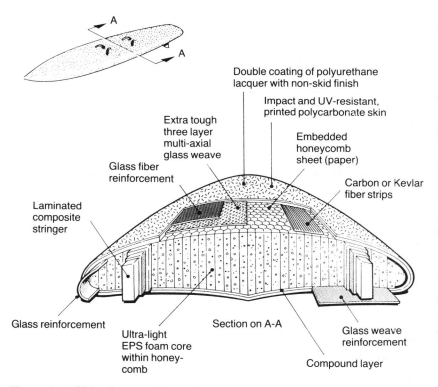

Figure 6.10 This schematic illustration of the cross-section of a surfboard shows it to be one of the most sophisticated of all sports equipment. From *Tomorrow's Materials* by K. E. Easterling (Institute of Materials, 1990).

whole is enclosed within a glass-fibre-reinforced epoxy composite, with extra fibre strengthening of Kevlar in parts likely to be exposed to extra wear and tear. Clearly, surfing boards of this complexity can only be produced by specialist companies. However, some enthusiasts like to build their own boards. Kits can be bought from which boards are constructed using extruded foam polystyrene (EPS) as a core material. This is then covered with glass fibre (or Kevlar fibre) weave, built up and shaped with resin. Higher-strength Kevlar or graphite weaves can be used instead of glass fibre, but in that case there is a ten times cost penalty. More recently, simpler surfboard constructions have been made by using a mouldable polyethylene copolymer. This foamed polymer reduces weight and avoids time-consuming hand-shaping operations. It should also be cheaper.

Currently, interesting research is in progress to improve the hydrodynamic properties of the underwater fins on surfboards and windsurfers in order to reduce turbulence. As with the America's Cup yachts, fluting hydrodynamic shape and high strength are all factors being investigated. The rise of carbon fibre/Kevlar polymer composites is vital in these highly stressed fins.

Sandwich

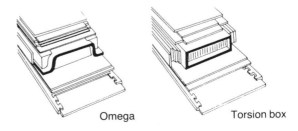

Omega   Torsion box

SYNOPSIS

The last decade has witnessed a large increase in the production of skis and associated equipment. Nowadays, over 15 million people per year, worldwide, find their way to downhill slopes and cross-country ski-trails. The majority of these skiers bear the last word in modern materials technology in the form of their skis, sticks and boots. The advanced materials selected reflect the need to balance performance with protection against injury. Indeed, skis are now so delicately constructed that it is claimed that the slightest changes in terrain or snow conditions can be 'felt' by the skier.

INTRODUCTION AND DESIGN PHILOSOPHY

Skiing is a highly dynamic sport with several forces acting on the skis simultaneously (Figure 7.1). On this basis, skis should ideally be selected according to the skier's weight, height and, of course, ability to handle the slopes. In other words, the type of ski chosen

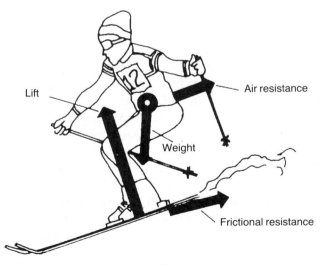

Figure 7.1 Forces acting on a skier. From *Tomorrow's Materials* by K. E. Easterling (The Institute of Materials, 1990).

Figure 7.2 Very high pressures are exerted on the ski edges in downhill skiing, and this is an important consideration in ski design and choice of material at the edges. From *Athletic Ability and the Anatomy of Motion* by R. Wirhead (Wolfe Medical Publications, 1989).

ought to reflect closely the skier's build and level of ability. If the skier is too heavy for the type of ski chosen, it will cause 'chatter' and poor edge control, especially under hard or icy snow conditions. Underloading, on the other hand, leads to difficulties in trim imitation, particularly in soft or deep snow.

In downhill skiing, the pressure on the skis resulting from the speed of movement and body weight (Figure 7.2) is constantly shifted from one side of the ski to the other, thus exposing the skis to high forces at their edges. Obviously, this pressure also varies as a function of the skier's speed, radius of turn, as well as the terrain conditions.

As a result of all these different forces and dynamics, skis have to be constructed to account for somewhat contradictory functions. On the one hand, the ski needs to provide for good longitudinal torsional rigidity to ensure the weight/pressure distribution is correct; on the other hand it should be flexible enough (Figure 7.3) to respond to different surface properties in order to dampen dangerous vibrations which may easily lead to hip and knee injuries.

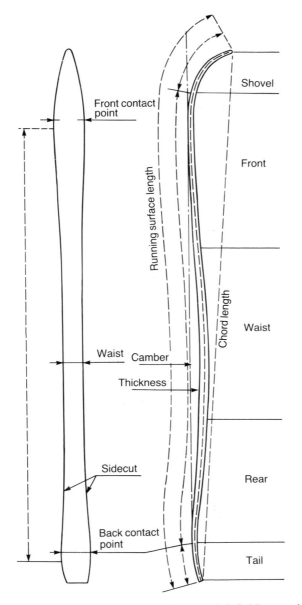

Figure 7.3 Schematic illustration of the shape and definitions assigned to the various parts of a ski. Note the allowance for ski flexibility at the centre from the skier's weight. From lecture notes of Dr Hugh Casey, Los Alamos National Laboratory.

It seems, therefore, logical that manufacturers have tended to develop two different categories of skis with respect to downhill skiing. A soft flexing ski which is used in soft or deep snow conditions, and a stiff rigid racing ski designed more for competition skiing. Recent developments, however, have shown that the different stresses exposed to skis in either category may be met by constructions using special composite designs, and that these skis can actually do well in both categories. Cross-country skis fall into a category of their own, of course, in terms of design and use of materials (Figure 7.4). Here, the emphasis is on extreme lightness and flexibility and this is again achieved by the use of cellular/composite materials. However, the demands even on these types of ski have increased in recent years with the development of the so-called 'skating' technique (Figure 7.5). This new technique, now acceptable in certain categories of international competitions, requires more attention to the design of ski edges; it also makes the skier more susceptible to injury due to the extra torsional stresses on the legs.

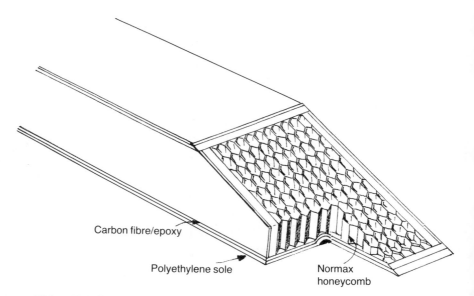

Carbon fibre/epoxy

Polyethylene sole

Normax honeycomb

Figure 7.4 Cross-section of a cross-country ski. These skis are designed to be extremely light, yet they must also be robust when used over tens of kilometres of mountain track. As shown in the illustration, a Nomax honeycomb core is bound by carbon fibre/epoxy sheets, with a polyethylene sole.

Figure 7.5 The 'skating' technique in cross-country skiing. Introduced in recent years by Gunde Svan of Sweden, this technique is now allowed in certain Olympic competitions. It differs from classical cross-country skiing in which the skis are held on parallel tracks. The skating technique is faster than classical skiing, but is more demanding on the skier and the ski edges.

DEVELOPMENTS IN SKI DESIGN

Developments in ski construction read in a similar way to many other high-tech sporting goods. As with bicycles and rackets, skis too began as monolithic wooden constructions. It was also customary in the early days for skiers to use a long pole (as a rather ineffectual brake) rather than hold sticks in each hand. The first step towards significant improvement in materials construction was brought about by the invention of laminated wooden skis in the late 1930s. This principle of connecting layers (or composites) of wood lamellae together in order to achieve superior toughness and strength properties is still used today, except that nowadays more advanced materials are contained in the composite. As with many other sporting goods, different new materials, brought about both by improved properties and cost-effectiveness in production, gradually changed and optimized performance characteristics. Steel and aluminium alloy box frames began to form the integral part of downhill ski manufacture and provide the necessary stiffness, rigidity and stability to the skis.

ADVANCED MATERIALS IN SKIS

It was in the 1960s that fibre-reinforced plastic was first introduced into ski construction as a means of improving durability, rigidity and (not least) cost-effectiveness. Since then FRP has been widely used in ski constructions, either in the form of pre-impregnated semi-cured sections, or as a laminate wrapped around a suitable core material. The principal advantage of this material lay in its ability to undergo extensive elastic strain without permanent (plastic) deformation occurring.

Basically three types of frame design are used in downhill skis: **torsion box, sandwich**, and **omega** constructions (Figure 7.6). Skis today are variations of all of these three basic types, and, according to their design, each tends to possess certain unique properties.

In general, sandwich constructions (a classical design) allow good stability along the ski's longitudinal axis, but are less appropriate when torsional forces are present. Sandwiched or laminated skis are produced by building up laminates of glass or carbon fibres (sometimes even glass–carbon hybrids) above and below a central core. These laminates are mated with an epoxy resin and pressure glued together.

In contrast to the classical sandwich design, the torsion box frame consists of a closed box within the ski cross-section, thereby providing excellent torsional rigidity. These skis are produced by wrapping layers of fibres around a central core, the

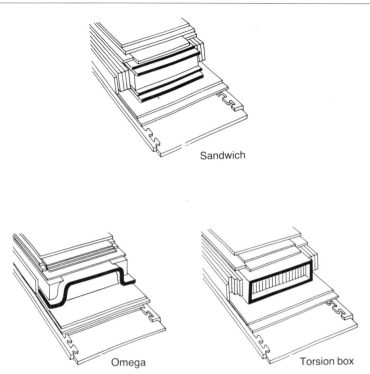

Figure 7.6 Three possible constructions of a downhill ski. In practice, combinations of these basic designs are used. From lecture notes of Dr Hugh Casey, Los Alamos National Laboratory.

edges and base being added afterwards. The whole assembly is then placed in a heated mould for curing.

The omega frame tries to combine the advantages of both the aforementioned systems. In particular, edge control is thought to be improved here, a definite bonus in competition slalom skiing for instance. The latest ski design attempts to obtain the best of all of these various designs. For example, skis with two lateral torsion boxes adjacent to an omega rib, all sandwiched between fibre-reinforced laminates, have now been produced in 'optimum' performance skis.

Core materials

All downhill and cross-country skis require a suitable core, invariably in the form of a cellular material. This can be wood (ash,

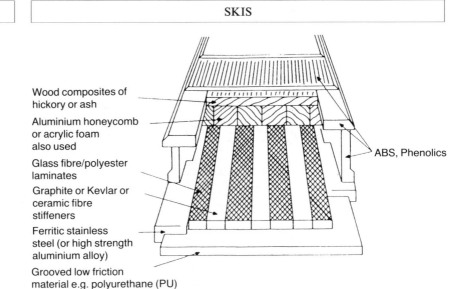

Wood composites of hickory or ash

Aluminium honeycomb or acrylic foam also used

Glass fibre/polyester laminates

Graphite or Kevlar or ceramic fibre stiffeners

Ferritic stainless steel (or high strength aluminium alloy)

Grooved low friction material e.g. polyurethane (PU)

ABS, Phenolics

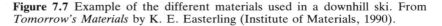

Figure 7.7 Example of the different materials used in a downhill ski. From *Tomorrow's Materials* by K. E. Easterling (Institute of Materials, 1990).

hickory, or African hardwood), although this material tends to carry weight penalties. A good alternative cellular material is in the form of a bee's honeycomb (see Chapter 2) and this may be made from aluminium or titanium foil in slalom skis. In cross-country skis, a paper honeycomb is often employed. Polyurethane foams have also been tried, although additional glass fibre reinforcement is then needed to improve strength and resilience of the core. Another advantage of cellular core material lies in its excellent shock-damping properties, a vital constituent in mountain skiing. More advanced skis may use acrylic foams which are even lighter than polyurethane, but these also tend to be more expensive.

Downhill skis typically consist of about 30 different parts, built up as a multi-layer laminate construction. Not surprisingly, advanced materials such as Kevlar, aramid and carbon fibres (Figure 7.7) are being used more and more in today's ski constructions. These materials are incorporated in the ski as layers around, or may even form, the core itself. For example, special lightweight skis have recently been made from an aramid honeycomb core, with layers of unidirectional and bias-ply fibre glass laminates, the main elements enclosing the core. This bias-ply

layer construction has been found to improve the ski's performance in harsh (icy) snow conditions. The top layers made from
ABS are protected with polyacrylether edges with the bottom
edges being made from an alloy steel.

The Kevlar aramid fibres are mainly included to improve the
ski's damping or shock-absorbing properties, this material being
somewhat better than carbon fibres in this respect. However,
carbon-fibre-reinforced epoxy is used for the majority of skis
nowadays, both downhill and track. Carbon fibres have outstand-
ing damping properties and tend (it is claimed) to create a 'more
spirited ski' with which skiers 'feel' the snow surface contours
better than when other composites are used.

TESTING SKI PROPERTIES

Major manufacturers need to carry out a number of comparative
tests on the dynamic behaviour of different ski constructions in
order to try to optimize their products with respect to their
handling properties. A particular concern is to optimize the
properties of different fibre types with respect to the best overall
lightness, toughness and dampening properties.

As an example of these tests, the core material of one test ski
consisted of polyurethane reinforced in glass fibre, or a honeycomb structure of meta-aramid fibre paper. The proportion of
fibre glass and aramid fibre in the laminate construction was
systematically varied, and the resonant frequency of the ski was
then measured, this being an indicator of the overall stiffness and
toughness of the ski. The results indicated, as may be expected,
a proportional increase in shock absorption properties as a function of increasing volume fraction of aramid fibres. The best
results were apparently achieved in this test when 35% aramid
fibres/65% fibre glass were used in conjunction with a core of
polyurethane (PU).

SAFETY EQUIPMENT IN SKIING

Skiing is certainly a sport prone to many types of injuries. Indeed,
one survey has shown that over a million people are injured on

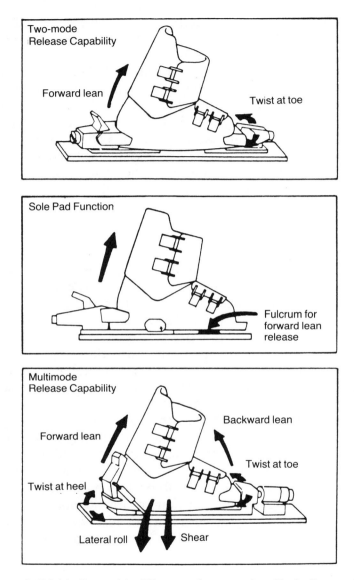

Figure 7.8 Ski bindings with different release modes. Typically copolymers (POM) or nylon are used as binding materials although there are many other alternative materials. From T. Reilly and A. Lees (1984) Exercise and sports equipment: some ergonomic aspects, *Applied Ergonomics*, December, p.259.

the ski slopes every year! Of these, about 40% to 50% are injuries to the lower extremities (feet, ankles) although this figure has declined from a peak of 64% a decade ago. This reduction is thought to be due to the development of better boots and bindings, rather than to any improvement in our skiing ability!

The binding is certainly a key element in the overall safety system (Figure 7.8). Its function is to hold the boot firmly to the ski, yet release it automatically if torsional loads between the ski and the boot become critical. It is 'critical' loading which is quite difficult to account for or predict. Most injuries in downhill skiing result from a failure of this binding system. There are a number of different release modes on the market and problems that need to be addressed include protection against backward falling movements, twisting movements and the release of the boot even if loading conditions are varying between these extremes. Some of these problems have been solved in modern bindings by including specially designed low-friction pads or rolls between the boot sole and the upper ski surface. More advanced electronic ski bindings are also now being tested.

MATERIALS IN BOOTS AND BINDINGS

Bindings

The first bindings were made from aluminium, but today most models rely on polymers with the largest proportion being from polyacetal, nylon, PU, ABS, high-density polyethylene and poly-tetrafluoroethene (PTFE). Of these, nylon resin has proved to have the best overall fatigue and abrasion resistance properties, and it also maintains its properties down to low temperatures, as is indeed required in this sport. Copolymers (POM) are now also being tried and these are found to possess good resilience and rigidity, besides being lighter than nylon.

Ski-boots

As all downhill skiers know, the boots are an important integral part of their skis. Ideally, therefore, the boots need to be fairly rigid, yet compact and light. Moulded PU or similar hard plastics are mainly used to achieve this, and nowadays 55% of ski-boot production is based on PU. It is in fact found that up to five different moulds may be needed to produce a single boot. More refinement in boot manufacture can be achieved, however, by injection moulding the outer parts of the boot from different grades of polymers. Boots made in this way effectively allow the skier more control over his or her movements as a result of the improved boot flexibility achieved. Even walking (between skiing) may become easier with this new design (ski-boots are always a problem in this respect because of their stiffness).

Alternative materials being tried for boots include surlyn ionomers and polyester elastomers. Surlyn ionomers are lighter than PU and nylon and their rigidity ranges less at low temperatures. In addition, from the manufacturers' point of view, polyester elastomers ought to receive more attention because of their high-speed processing capabilities.

SKI-STICKS

As with the other ski equipment ski-sticks, too, have undergone continual and radical developments. Ideally, ski-sticks should combine minimum weight with maximum rigidity and good impact resistance. For this purpose carbon-fibre-reinforced plastics are now used, and these are made in a similar way to those of racing-yacht masts, using a special fibre-winding machine. Three or four layers of fabrics made from carbon and glass fibre weavings are attached as a sheath to the torsion resistant windings, orientated at 0° and 90° to the axis of the stick. In order to reduce weight, the wall thickness of a pole may normally not exceed 0.5–0.6 mm, although this thin shell is obviously susceptible to impact problems.

Further reading

The data collected in this book derive from a large number of sources, including brochures, newspapers, magazines, journals and books. Of the books, the following were found particularly useful:

Brandt, J. (1983) *The Bicycle Wheel*, Avocet, London. All you ever need to know about the history and detailed construction of the spoked bicycle wheel.

Cotterhill, R. (1985) *Cambridge Guide to the Material World*, Cambridge University Press. A beautifully illustrated text covering all types of natural and synthetic materials.

Easterling, K. E. (1990) *Tomorrow's Materials*, 2nd ed., The Institute of Materials, London. An introduction to materials science with modern applications including sporting materials.

Gordon, J. E. (1976) *The New Science of Strong Materials*, Penguin, Harmondsworth. An excellent and often amusing survey of materials science and its applications.

Gordon, J. E. (1978) *Structures*, Penguin, Harmondsworth. On the same lines as the previous book; a good read, amusing and informative.

Nigg, B. N. ed. (1986) *Biomechanics of Running Shoes*, Human Kinetics Publishers, London. An impressive, fairly scientific survey of the testing, design and construction of sports shoes.

Peterson, L. and Renstrom P. (1986) *Sports Injuries*, Martin Dunitz, New York, London. This is an elementary, easy-to-read introduction to the subject and covers most types of injuries sustained in sporting activities.

Sayer, Bill (1991) *Rowing and Sculling*, Robert Hale, London. A really excellent overview of rowing, rowboats and rowboat construction.

Whitt, F. R. and Wilson, D. G. (1982) *Bicycling Science*, MIT Press, Cambridge, MA. A good survey of the history, science and technology of bicycles and their detailed construction.

Wirhed, R. (1989) *Athletic Ability and the Anatomy of Motion*, Wolfe Medical Publications, London. A beautifully illustrated text on human motion in sport as seen from an anatomical and physiological point of view.

Index